The Planet Venus

also by Patrick Moore

YOUR BOOK OF ASTRONOMY

THE PLANET JUPITER *by Bertrand M. Peek,*
revised by Patrick Moore

One of a sequence of nine ultra-violet images of Venus,
obtained by the Pioneer Venus Orbiter Imaging System
on 16 February 1979.

The Planet Venus

Garry E. Hunt and Patrick Moore

faber and faber

First published in 1982
by Faber and Faber Limited
3 Queen Square London WC1N 3AU
Filmset by Wilmaset, Birkenhead
Printed in Great Britain by
Ebenezer Baylis and Son, Limited
The Trinity Press, Worcester and London

© *Garry E. Hunt and Patrick Moore, 1982*

British Library Cataloguing in Publication Data

Hunt, Garry
   The planet Venus.
   1. Venus (Planet)
   I. Title   II. Moore, Patrick
   523.4′2   QB621

   ISBN 0-571-09050-8

Library of Congress Cataloging in Publication Data

Hunt, Garry E.
   The planet Venus.

Includes index.
   1. Venus (Planet)   I. Moore, Patrick.   II. Title.
   QB621.H86   523.4′2   82-5045
   ISBN 0-517-09050-8   AACR2

Acknowledgements

Our thanks for permission to reproduce the illustrations are due to the following. Plate 1: Royal Observatory, Edinburgh; plate 5: Mount Wilson and Palomar Observatories; plate 6: International Planetary Patrol Programme, Arizona; plates 10, 13, 15, 16 and figures 15–32: Journal of Geophysical Research, American Geophysical Union; plates 11, 12, 14, 16: NASA and the Jet Propulsion Laboratory. We are also grateful to Paul Doherty for his skilful line drawings, and to the publishers, in particular Charles Monteith.

G.E.H.
P.M.

Contents

Illustrations

Foreword

The idea for the present book was put forward some years ago, when it became clear that Patrick Moore's original small book about Venus, originally published in 1956 and last revised in 1963, was completely out of date. The present book is therefore totally new. It is, of course, the result of the closest collaboration between the two authors. Basically, Moore wrote Part 1 and the first part of Chapter 10, 12 and 14, while Garry Hunt wrote Part 2, but everything has been discussed in great detail between us.

We hope that we have presented a balanced picture of Venus as we now know it. A remarkable amount of information has been collected during the past two decades; Venus is not the kind of world which we had expected it to be. No doubt there will be further surprises in the future. Meanwhile, we have made this book as complete as possible.

G.E.H.

March 1982

P.M.

Part 1 Venus before the Space Age

1 The 'Evening Star'

The planet Venus is our near neighbour in space. Among natural bodies only the Moon is closer to us, and it is not surprising that Venus shines so brilliantly in our skies.

What makes it particularly interesting is the fact that in some ways it seems to be almost a twin of the Earth. It is only slightly smaller and less massive, and on average it is a mere 41,000,000 kilometres closer to the Sun, so that there is no obvious reason why it should not be in a similar state. Before the era of space-probes there were conflicting views about it, but although it was presumably a hot world there were many astronomers who believed it to be ocean-covered, in which case the existence of life could not be ruled out. Only during the past two decades have we discovered that Venus is probably the most hostile world in the entire Solar System.

Five planets—Mercury, Venus, Mars, Jupiter and Saturn—have been known from very early times. The philosophers of Ancient Greece were well aware of their special nature, and no doubt their wanderings against the starry background were known even before then. Mars, Jupiter and Saturn move around, keeping within a definite belt in the sky known as the Zodiac. Mercury and Venus keep comparatively close to the Sun, and it was reasoned that they must lie closer to us than the Sun. We now know that they move in orbits well inside that of the Earth, so that they are called the Inferior Planets.

Mercury is never very prominent, but Venus cannot be mistaken, and is far more brilliant than any other planet or star. It can never be seen all through any one night, but as an evening object it may not set until several hours after the Sun, and at morning apparitions it may rise while the sky is still dark.

The Chinese called Venus Tai-pe, or 'the Beautiful White One'.[1] In Egypt, Venus or 'Bonou', the Bird, was Ouâiti as an evening star and Tioumoutiri as a morning star.[2] Originally it was no doubt believed that the evening star and the morning star were two different bodies, but systematic observations showed conclusively that they must be one and the same.

The Babylonians called Venus 'Istar', the personification of woman and the mother of the gods, and described her as 'the bright torch of heaven'.[3] Temples to her were set up in Nineveh and various other places. Istar was regarded as being responsible for the world's fertility, and there is a legend that when she visited the Underworld to search for her dead lover Tammuz all life on Earth began to die—to be saved only by the intervention of the gods, who revived Tammuz and so restored Istar to the world. There is an obvious similarity here with the famous Greek legend of Demeter and Persephone (or, to use their more familiar names, Ceres and Proserpina). It is worth noting that the first observations of Venus on record come from Babylonia,[4] and are found on the famous Venus Tablet, discovered by Sir Henry Layard at Konyunjik and now in the British Museum.[5] The date is given as about 1700 BC,

Venus was generally regarded as female, except in India. The Greeks and Romans named the planet in honour of the Goddess of Beauty, and temples of Venus were set up in various places. The month of April was regarded as sacred to the goddess, and our name 'Friday' is derived from the Anglo-Saxon 'Frigedæg' (Friga, or Venus, and dæg, day). Venus-worship continued until surprisingly modern times. It has been claimed that in Polynesia human sacrifices were offered to the Morning Star as lately as the nineteenth century,[6] and there is evidence that sacrifices were also offered by the Skidi Pawnee Indians of Nebraska.[7] Ancient beliefs take a long time to die.

The Egyptians, naturally enough, regarded the Earth as the centre of the universe, with all other bodies—including Venus—moving round it. (Suggestions that they regarded Mercury and Venus as satellites of the Sun[8] seem to rest on very uncertain evidence.) With the Greeks, we come to the beginning of astronomy as a true science. Homer refers to Venus as 'Hesperos, which is the most beautiful star set in the sky';[9] the great philosophers studied its movements carefully, and drew up tables which were at least reasonably accurate. A few far-sighted Greeks, notably Aristarchus of Samos, even dethroned our world from its proud position in the centre of the universe, and relegated the Earth to the status of a mere planet moving round the Sun. Aristarchus lived from around 310 to 250 BC, so that he anticipated Copernicus by eighteen centuries. Unfortunately he found few followers; the step was too bold for even the Greeks to take, and later philosophers reverted to the idea of a central, stationary Earth.

This scheme of things reached its highest perfection in the work of Hipparchus (*circa* 140 BC) and Ptolemy (*circa* AD 120–80). Ptolemy, or to give him his correct name Claudius Ptolemæus, produced a great book which is generally known as the *Almagest*.[10] The book is really a summary of the state of scientific knowledge at the end of Greek times, and the system of the universe accepted at that period is always called the Ptolemaic Theory, though Ptolemy himself did not invent it. Ptolemy is a somewhat shadowy figure; of his life and personality we know absolutely nothing, except that he lived and worked in Alexandria. Periodical attempts to discredit him, and to claim that he was a copyist at best, have not been notably successful; but in any case science owes him a debt of gratitude. His book has come down to us by way of its Arab translation. Without it, we would know much less about ancient science than we actually do.

The Ptolemaic Earth lies in the centre of the universe, with the various celestial bodies revolving round it in perfectly circular orbits; they had to be circular, because the circle is the 'perfect' form, and nothing short of perfection can be allowed in the heavens. First comes the Moon, the closest body in the sky; then Mercury, Venus and the Sun, followed by the three other planets then known (Mars, Jupiter and Saturn), and finally the sphere of the so-called fixed stars.

This may sound delightfully simple, but actually it was nothing of the kind, because the theory of planets moving at uniform speeds in circular orbits did not fit the observations—as Ptolemy knew quite well. For instance, the planets do not move steadily from west to east against the stars. Mars, Jupiter and Saturn regularly stand still briefly and then move in a backward or retrograde direction before resuming their eastward march. To overcome this problem, Ptolemy assumed that a planet moved in a small circle or epicycle, the centre of which—the deferent—itself moved round the Earth in a perfect circle. The possibility of elliptical orbits was not considered, so that there was really no other way out of the difficulty. Moreover, a single epicycle would not suffice; others had to be added, until the whole system had become hopelessly clumsy and artificial.

Mercury and Venus presented problems of their own, and Ptolemy was forced to assume that their deferents remained permanently in a straight line with the Sun and the Earth. At least this explained why Mercury and Venus never appear opposite to the Sun in the sky.

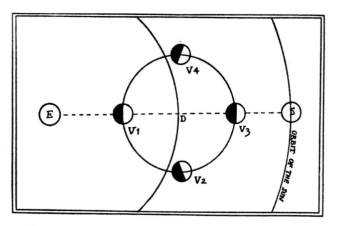

Fig. 1   Movements of Venus, according to the Ptolemaic system.

Fig. 1 shows the movement of Venus according to the Ptolemaic System. E is the Earth, at rest in the centre of the universe; S is the Sun, D is the deferent of Venus, and V1 to V4 the planet in four positions as it moves in its small circle or epicycle. The line EDS must always be straight, and there is one vitally important consequence. Since a planet shines only by reflected sunlight, it is obvious that on the Ptolemaic theory Venus can never be seen as a full disk, or even a half. At V1 and V3 the dark hemisphere will be turned towards us, so that Venus will be invisible; at V2 and V4, part of the sunlit side will face the Earth, and Venus will appear as a crescent.

No such changes of shape, or phases, can ordinarily be detected with the naked eye, and so the theory could not be disproved on observational grounds. There, for many centuries, the matter rested. The idea of a central Earth was so deeply ingrained that to question it was regarded as heretical. From the ninth to the fifteenth centuries AD the Arabs made careful studies of the movements of Venus, and the planet was also observed in the New World; the Maya used it as the basis for their calendar, and there have been suggestions that this dates back to pre-Arab times.[11] Then, in 1543, came the publication of the book by Copernicus which sparked off one of the greatest controversies in the whole history of science.[12]

Copernicus, or more properly Mikołaj Kopernik, was a Polish canon with an absorbing interest in astronomy. His main contribution was that he removed the Earth from the centre of the planetary system, and put the Sun there instead. To be honest, this was his only real triumph; in most other ways he was wrong. In particular, he retained the idea of circular orbits, and was even reduced to bringing back epicycles. Yet he had taken the decisive step, and the battle began, but Copernicus himself was not involved, because he had been wise enough to defer publication of his book until the last days of his life. As he had expected, the Church was implacably hostile to the new ideas, and some of Copernicus' followers were savagely persecuted; in 1600 Giordano Bruno was burned at the stake in Rome, one of his crimes being that he persisted in teaching the Copernican theory rather than the Ptolemaic.

Telescopes were invented in the early seventeenth century. (They may have been known rather earlier, but we have no proof.) In the winter of 1609–10 Galileo Galilei, Professor of Mathematics at Padua, built a telescope for himself and turned it skyward. By modern standards it was tiny, but it was enough to lead Galileo on to a series of dramatic discoveries. The Moon showed lofty mountains and vast craters; the Sun was not pure, but spotted; Jupiter was attended by four satellites, and there was something curious about the appearance of Saturn, though Galileo could not make out what it was. The Milky Way turned out to be made up of faint stars.[13]

Galileo was already a firm believer in the Copernican theory, and he looked for proof. He found it in the phases of Venus. There were phases indeed, but they were of the lunar type; sometimes Venus was a crescent, sometimes a half, sometimes almost full.

In those days it was often the custom to announce discoveries in anagram form. Accordingly, Galileo sent the following message to the great mathematical astronomer Johannes Kepler: '*Hæc immatura, a me, iam frustra, leguntur—o.y.*' In this form, the message may be translated as 'These things not ripe (for disclosure) are read by me'—the letters 'o.y.' are tacked on, since there is no way of fitting them into the original sentence. Rearranged, the letters give: '*Cynthiæ figuras annulatur Mater Amorum*', or 'the Mother of the Loves imitates the phases of Cynthia'.[14] The Mother of the Loves is, of course, Venus; Cynthia is the Moon. Galileo had been quick to realize the significance of what he had seen. Venus could not move in the way that Ptolemy had assumed, since according to the Ptolemaic theory it could never show up as a half or three-quarter disk.

The facts were undeniable, and Galileo made full use of them in his great work, known usually as the *Dialogue*.[15] Copernicus had been basically right, though he had made many errors in detail—such as suggesting that Venus might be either self-luminous or else transparent.[16]

Galileo's discoveries met with the usual stormy reception. This is no place to tell the story of his trial and condemnation; it is enough to say that his telescopic discoveries made a tremendous contribution to the downfall of the Ptolemaic theory.

Kepler was under no delusions. His studies, based on the accurate measurements of the movements of Mars made by the Danish astronomer Tycho Brahe, had enabled him to draw up the famous Laws of Planetary Motion which bear his name. The first of these Laws states that the planets move round the Sun in ellipses, the Sun occupying one of the foci of the ellipse; the behaviour of Venus was merely one of the proofs which he had been expecting. Before the end of the century, Isaac Newton's work upon universal gravitation had put the whole matter beyond any shadow of doubt.

The phases of Venus had helped to open the gateway to knowledge; the road ahead seemed clear.

2 Venus as a World

The principal planets of the Solar System are divided into two well-marked groups. Mercury, Venus, the Earth and Mars, which are solid and relatively small, make up the inner group. Beyond Mars we come to the minor planets or asteroids, which are midget worlds; only one of them (Ceres) is as much as 1000 kilometres in diameter, and there are very few with diameters of more than 300 kilometres. Only one member of the group, Vesta, is ever visible with the naked eye. Even if combined, all the asteroids would not make up one body as massive as the Moon. Beyond this zone we come to the four giants: Jupiter, Saturn, Uranus and Neptune, which have gaseous surfaces and are completely different from the inner planets. Finally there is Pluto, which is in every way a puzzle. It was discovered in 1930; it seems to be smaller than the Moon, and is of low density. Its orbit is relatively eccentric, and from 1979 to 1999 its distance from the Sun remains less than that of Neptune. There are now serious doubts as to whether Pluto is worthy of true planetary status.[1] The main planetary data are summarized in the table below.

Venus appears particularly brilliant not mainly because of its size, but because of its closeness and its high albedo or reflecting power. It reflects 76 per cent of the sunlight falling upon it, as against only 7 per cent on average for the Moon and 16 per cent for Mars.[2] This brightness is due to the fact that when we look at Venus, we are seeing not the actual surface, but simply the top of a layer of cloud. True, the Cytherean* clouds are different in character from those of Earth, but they are very efficient at reflecting the rays of the Sun.

In the evening or morning sky Venus is a superb object, glowing down almost like a small lamp, and even casting a shadow at times.

*There is no generally accepted adjective for Venus. 'Venusian' is common but ugly; 'Venerian' is even worse. 'Cytherean', an adjective derived from the old Sicilian name for Venus, is perhaps preferable, though not strictly correct, and will be used throughout this book.

Planet	Mean distance from Sun, millions km	Sidereal Period	Orbital Eccentricity	Orbital Inclination, °	Axial Rotation	Axial Inclination °'
Mercury	57·9	87.96d	0.206	7	58·56d	28
Venus	108·2	224·7d	0·007	3·4	243·01d	178
Earth	149·6	365·3d	0·017	0	23h 57m	23·27
Mars	227·9	687·0d	0·093	1·9	24h 37m	23·59
Jupiter	778·3	11·86y	0·048	1·3	9h 55m 29·7s	3·05
Saturn	1425·0	29·46y	0·056	2·5	10h 39m	26·44
Uranus	2867·0	84·01y	0·047	0·8	16·31±0·27h	97·52
Neptune	4496·6	164·79y	0·009	1·8	±18h	28·48
Pluto	5890·0	247·7y	0·248	17·2	6·37d	?

Planet	Equatorial Diameter, km	Surface Gravity, Earth=1	Density, Water=1	Mass, Earth=1	Volume, Earth=1	Mean Temperature, °C
Mercury	4878	0·37	5·4	0·07	0·06	+350 (day), −170 (night)
Venus	12,104	0·88	5·2	0·82	0·88	−33 (cloud), +480 (surface)
Earth	12,756	1	5·5	1	1	+22
Mars	6786	0·38	3·9	0·11	0·15	−23
Jupiter	142,796	2·64	1·3	318	1316	−150 (clouds)
Saturn	120,000	1·15	0·7	95	755	−180 (clouds)
Uranus	51,800	1·17	1·2	15	67	−210 (clouds)
Neptune	49,500	1·18	1·7	17	57	−220 (clouds)
Pluto	2440	?	1?	?	0·01?	−250?

Under favourable conditions the planet may be seen in broad daylight, provided that one knows just where to look. This was realized in ancient times; Varro records that 'in his voyage from Troy to Italy, Æneas constantly perceived this planet, notwithstanding the presence of the Sun above the horizon.' More recently, an instance of the daylight visibility of Venus has crept into history. The French astronomer François Arago gives the following account:[3]

'Bouvard has related to me that General Bonaparte, upon repairing to Luxembourg, when the Directory was about to give him a fête, was very much surprised at seeing the multitude which was collected in the Rue de Tournon pay more attention to the region of the heavens situate above the palace than to his person or to the brilliant staff which accompanied him. He inquired of the cause, and learned that these curious persons were observing with astonishment, although it was noon, a star, which they

1   Venus and Halley's Comet, 1910. Lowell Observatory, Flagstaff, Arizona.

supposed to be that of the Conqueror of Italy; an allusion to which the illustrious general did not seem indifferent when he himself with his piercing eyes remarked the radiant body. The star in question was no other than Venus.'

Napoleon presumably regarded the sign as one of divine approval, but Venus has at times caused alarm. In November 1887, for instance, it was a brilliant morning star, and a well-known British astronomer, Sir Norman Lockyer, felt it necessary to write an article in which he explained the reasons for the planet's unusual brightness,[4] at the same time giving his opinion that there was an urgent need for more science teaching in board schools.[5] In 1916, some people in England mistook Venus and Jupiter, which happened to lie close together in the sky, for the fore and rear lights of a Zeppelin.[6] More recently, Admiral A. J. L. Murray has related an experience which is not without its humorous side.[7] On 15 November 1939, shortly after the outbreak of war, HMS *Cornwall*, on active service, received a message from a sloop on passage from Ceylon to Aden that star-shell firing had been seen on a bearing of 247°. Preparations were made, but nothing further happened. Next day Admiral Murray in the *Cornwall* worked out the bearing of Venus as it set on that night—247°....

And, of course, Venus has been responsible for innumerable flying saucer reports. There seems very little doubt, for instance, that this was the explanation of the Unidentified Flying Object described by President Carter of the United States!

Shadows cast by the planet have also been recorded since ancient times. Pliny noticed them,[8] and so did the Greek astronomer Simplicius.[9] In more modern times they have been mentioned by Sir John Herschel, and by E. M. Antoniadi in 1897.[10] One of the present writers (Moore) has seen a clear shadow cast by Venus when near maximum brilliancy.

In 1876 J. I. Plummer[11] threw shadows from Venus on to a screen and compared them with shadows cast by a candle at varying distance, and in 1956 some interesting comments were made by W. H. Steavenson.[12] Steavenson pointed out that the shadows cast by Venus are very sharp, since the planet is practically a point source of light, and the penumbral effects associated with shadows cast by the Sun and Moon are virtually absent. He wrote:

'But in spite of this sharpness of definition, the shadows are at best rather delicate objects when observed in the open air. This is because the general illumination of the sky ... floods the receiving surface with faint light, and so impairs the contrast.

'It is obvious that much of this general illumination can be eliminated by making the observation in a room into which only the light of Venus, plus a few degrees of the surrounding sky, is admitted through a small window. Under these conditions the shadow of the window-frame stands out quite strikingly on the opposite wall, as has often been observed.

'The interior of an observatory may also be a suitable place for the display of the shadows. ... On the evening of 28 April this year, when Venus was about 15° above the horizon, I noticed that this white surface (the internal surface of the dome) was covered with a faint reticulated pattern of narrow dusky lines. A glance at Venus showed that the planet was passing behind the branches of a nearby leafless tree, and it was evident that the pattern was produced by the shadows of the individual twigs ... they could be seen in quite rapid motion. A week later some leaves had appeared on the tree, and the shadows of these could also be seen separately as dark blotches among the twigs.'

Another interesting phenomenon to be seen now and then is the so-called Green Flash. At the moment of its final disappearance below the horizon, the upper limb of the Sun may turn green for an instant, and this has also been seen in the case of Venus.[13] The flash was well seen by Admiral Murray from HMS *Cornwall*, off Colombo, at 13.50 (one and a half hours after sunset) on 28 November 1939, and was described by him as 'emerald'.[14] Venus was then setting over a sea horizon. Admiral Murray was using binoculars, and is doubtful whether he would have noticed the flash had he not been deliberately looking for it. It is, of course, an optical phenomenon due to the atmosphere of the Earth.

Venus was once believed to be slightly larger than the Earth,[15] but this is now known to be wrong; it is slightly but appreciably smaller. In 1900 T. J. J. See made a series of measurements with the 66-cm refractor at Washington in the United States, and gave a value of 12,164 kilometres,[16] which has proved to be very near the truth. Obviously, what we are measuring is the diameter of the upper cloud surface, not the solid globe itself, and the atmosphere of Venus is of appreciable thickness. In 1953 A. A. Nefedjev reported slight variations in diameter,[17] but these have not been confirmed, and they would be very difficult to explain. In 1964 the French astronomer G. de Vaucouleurs examined all the measurements, and gave a value of 12,240 kilometres, with an uncertainty of 15 kilometres either way.[18] The values now adopted, from radar measurements, are 12,240 kilometres for the cloud-top diameter, and 12,102 kilometres for the actual solid surface. The polar flattening is inappreciable. During the last century Vidal[19] and Tennant[20] reported measurable flattening, but it now seems that Venus is to all intents and purposes a perfect sphere—which, in view of its very slow rotation, is not surprising.

The question of the rotation period will be discussed in detail later in the present book. For the moment, suffice it to say that it is just over 243 Earth-days, which is longer than the Cytherean 'year' of just under 225 Earth-days—a case unique in the Solar System. Moreover, the rotation is in a retrograde sense: east to west instead of west to east, as with the Earth. This means that the length of the 'solar day' is 116·8 Earth-days, and if the Sun could be seen from the surface of Venus it would rise in a westerly direction and set towards the east.

The mean density of Venus is 5·2 times that of water, rather less than for Mercury or the Earth but appreciably greater than for Mars. It follows that the heavy core of the planet is smaller than that of the Earth, and it is not surprising to find that there is no measurable general magnetic field. The mass is 0·88 that of the Earth. Yet the differences between Venus and our world are relatively slight; it has been said that if the Earth were reduced in size and weight to that of a billiard-ball, Venus would be another billiard-ball so like the first that it would still be usable for play. The surface gravity, too, is only a little lower than ours. An Earthman standing on Venus would find that he weighed much the same as at home; there would be no sensation of unnatural lightness, as on the Moon or even on Mars.

Mass is linked with escape velocity—that is to say, the velocity which must be imparted to a body to enable it to break free without any extra impetus. In the case of the Earth, the value is 11·2 km/second (in Imperial units, 7 miles per second). With Venus, escape velocity is 10·4 km/second. It might therefore be thought that the atmosphere would be of about the same extent and density as our own, but this is not so. Instead of being oxygen-rich and breathable, the atmosphere of Venus contains a high proportion of the heavy gas carbon dioxide, and on the surface the atmospheric pressure is over 90 times that of the Earth's air at sea-level. This is certainly curious, and we have to do our best to find a good reason.

There can be little doubt that the root cause is the lesser distance of Venus from the Sun; the temperature has always been higher, and this has affected the whole course of the planet's evolution. If the Earth had been slightly closer to the Sun than it actually is, it could well have developed in the way that Venus has done—so that you and I would not be here.

Venus, then, is a world which may be a near-twin of the Earth in size, mass and escape velocity, but is in a very different condition. It is equally unlike the heavily cratered Mercury, which has only a trace of atmosphere, and neither does it resemble Mars. In almost every way it seems to be in a category of its own, and in view of its high surface temperature, its crushing atmospheric pressure and its unpleasant clouds, it must be regarded as something of a disappointment. Before the Space Age, Venus was widely regarded as a potential colony. Now, we have to resign ourselves to the fact that it is as hostile as it could possibly be. This does not make it any the less fascinating, but astronauts will not visit Venus for a long time in the future—if ever.

3 The Movements of Venus

Venus, as we have noted, is the second planet reckoning outwards from the Sun. Its orbit has the low eccentricity of 0·007, and is thus practically circular, more so than for any other planet (its nearest rival is Neptune, with 0·009). On average, Venus receives about twice as much solar radiation as we do. To us, the Sun has an apparent diameter of about 32 minutes of arc, but from Venus this would be increased to over 44 minutes.

Although the orbit of Venus is so nearly circular, it is tilted with respect to that of the Earth. The angle of inclination is 3° 24′. This may not seem a great deal, but it is more than for any other planet apart from Mercury (not counting Pluto, whose planetary status is now seriously in doubt). One result of this is that transits of Venus across the Sun's disk are few and far between.

Since Venus is closer to the Sun than we are, it moves more quickly. The Earth's mean orbital velocity is 29·8 km/second; that of Venus is 35·0 km/second. Travelling at this greater rate in a smaller orbit, Venus takes less time to complete one circuit. The Cytherean 'year' or sidereal period is only 224 days 16 hours 48 minutes.

The movements of Venus have been under investigation for a long time.[1] The cause of the phases is obvious enough. The diagram in Fig. 2 should explain it at once. In the position marked Inferior Conjunction Venus has its unlit or night side turned towards us, and is 'new'; it then appears very close to the Sun in the sky, and is to all intents and purposes out of view except on the rare occasions when the alignment is perfect, so that Venus shows as a black disk silhouetted against the Sun. At Superior Conjunction Venus is full, since its day side is facing us, but is almost behind the Sun. When at the two positions marked Elongation, Venus reaches its greatest angular distance from the Sun—about 47°—and is a magnificent object. At eastern elongation it is an evening star, narrowing and approaching 'new', while at western elongation it is a morning star waxing towards 'full'.

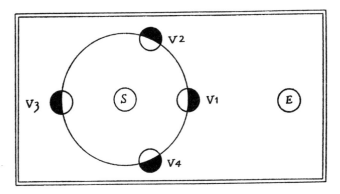

Fig. 2   The Phases of Venus.
V1 = Inferior Conjunction; V2, V4 = Elongation; V3 = Superior Conjunction.

The synodic period of Venus, or average interval between successive inferior conjunctions, is 584 days, though this may vary by as much as four days either way. In general, about 144 days elapse between evening and morning elongation, while 440 days are required for the planet to go round the Sun and return again to evening elongation. A list of future conjunctions and elongations is given in Appendix 3.

In view of the fact that Galileo's discovery of the phases of Venus took 'conventional' astronomers by surprise, it is reasonable to suppose that the phases themselves are quite invisible without binoculars or a telescope. Certainly the normal eye cannot see them, but it seems that they can be made out by people with exceptional sight under very favourable conditions. It is also interesting to note that ancient legends often refer to the horns of Venus, which some authorities have taken to indicate a crescent. Pliny represents Venus as a human figure with two horns;[2] the people of Samoa also said that Venus is horned,[3] and so on. One of the discoveries made by Layard was that of Astarte, the Assyrian Venus, bearing a staff tipped with a crescent.[4] Opinions as to the significance of these records differ. For instance, J. Offord stated that 'the universal acceptance of the crescent symbol can be accounted for only by the crescent form of Venus having been observed',[5] while W. W. Campbell wrote that he was 'inclined to attribute the ancient description of Venus as a crescent to a pure lucky guess, probably made under the influence of a crescent moon'.[6] Considerable doubt must remain.

H. McEwen, for many years Director of the Mercury and Venus Section of the British Astronomical Association, stated in 1895 that although the crescent might be seen with the unaided eye, this was very exceptional;[7] later he changed his mind, and wrote that the crescent was 'far beyond' naked-eye visibility.[8] On the other hand R. A. Proctor, a well-known British astronomer who wrote many books during the latter part of the last century, held that 'we know that under favourable circumstances we can ourselves recognize the crescent form of Venus with the unaided eye'.[9] The question is not important, and has of course nothing directly to do with Venus as a world; nevertheless, it is interesting enough to consider in slightly more detail.

There are many cases on record of the naked-eye visibility of the crescent. In the clear skies of South America, Lieutenant Gilliss recorded it on various occasions between 1849 and 1852.[10] Between 1929 and 1935 it was also recorded unmistakably by Carl Reinhardt,[11] D. Howell,[12] H. W. Cornell[13] and Dr and Mrs F. W. Wood.[14] Miss M. A. Blagg, well-known for her work in connection with the Moon, was unable to make out the crescent shape, but could see that Venus was definitely elongated.[15] The Rev T. W. Webb relates that the crescent phase was seen by a twelve-year-old boy, Theodore Parker, before he knew of its existence,[16] while W. S. Franks, a winner of the Gold Medal of the Royal Astronomical Society, said that his son, E. S. Franks, had frequently seen the crescent between 1890 and 1900.[17]

All these cases are well authenticated, and there seems therefore little doubt that the phase really is visible to people with exceptional eyesight. On the other hand, such observations are very difficult, and it is easy to deceive oneself, as one of the present writers (Moore) proved in 1957 by a simple though rather unkind experiment. Using a small 7-cm telescope, the crescent Venus was shown to eight people, whose ages ranged between sixteen and sixty and none of whom knew a great deal about astronomy. They were then asked whether they could see the crescent without optical aid, the six of them answered 'yes', while the other two were not sure. When given paper and pencil and asked to draw the crescent, all six claimants put down the crescent with the horns pointing to the west, as in the telescopic view. Naturally they did not realize that an astronomical telescope gives an inverted image, so that to the unaided eye the crescent will face in the opposite direction!

A rather more significant experiment was carried out by Moore some years later. During one of the BBC *Sky at Night* television programmes, a telescopic view of the crescent Venus was shown, and viewers who believed they could see the phase were asked to send in drawings. More than two hundred were received. Again, almost all of them showed the telescopic view, but there were two writers—both young—who said that, to their surprise, they could see the crescent as opposite to the view shown on the television screen. Therefore, it seems that these two people genuinely saw the crescent phase.

The minimum distance between the Earth and Venus occurs at inferior conjunction, and may then be reduced to about 39,000,000 kilometres, which is some 16,000,000 kilometres closer than Mars at its nearest. Unfortunately, Venus has then its night side facing us; and as the phase grows, the distance also increases, so that the apparent diameter of the disk shrinks. This is shown in Fig. 3. The black circle represents the size at inferior conjunction, when the diameter is 65 seconds of arc; the third position shows the apparent size at the moment of greatest brilliancy (angular distance from the Sun about 40°), and the last white disk represents the size at superior conjunction, when the apparent angular diameter has been reduced to a mere $9\frac{1}{2}$ seconds of arc, smaller than that of remote Saturn.

Fig. 3
Apparent size of Venus
at various phases.

At first sight it is surprising to find that Venus is at its most brilliant during the crescent stage, but the increasing phase has to be balanced against the shrinking angular diameter, as was pointed out long ago by Edmond Halley, the second Astronomer Royal.[18] Another problem is connected with the moment of exact half-phase or 'dichotomy'. The word dichotomy comes from the Greek, and means 'cut in half', so that Venus then appears as though sliced in two—one side being visible and the other invisible. Since the movements of the planet are known with great accuracy, it should be possible to forecast the moment of dichotomy very precisely. Actually, the predictions are often several days in error. The cause of this curious discrepancy seems to be linked with the Cytherean atmosphere, and will be discussed below.

It is clear that there is a difference between the 'limb' of Venus, which is merely the edge of the visible disk as seen from Earth, and the 'terminator', which is the actual boundary between the day and night hemispheres of the planet. In Fig. 4 the limb is shown as a continuous line, while the terminator is dotted.

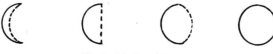

Fig. 4   Limb and Terminator.

Before turning to other matters, it is worth saying something about the strange theories put forward in 1950 by a Russian-born psychoanalyst, Immanuel Velikovsky, who practised in what is now Israel before emigrating to the United States in 1939. In his book *Worlds in Collision*,[19] he claimed that Venus used to be a comet, that it was ejected from Jupiter, and subsequently made various close approaches to the Earth, causing events which are linked with references in the Bible. In 1500 BC, for instance, at the time of the Israelite Exodus, the comet Venus caused a temporary halt in the Earth's spin, so that the Red Sea was left high and dry for long enough for the Israelites to cross. Conveniently, the rotation started up again just in time to swallow up the pursuing Egyptians. The comet Venus returned later on, producing thunder, lightning and other effects noted when Moses was given the Ten Commandments on Mount Sinai. Later encounters produced new phenomena, such as the shaking-down of the walls of Jericho. Finally the comet Venus collided with Mars, and had its tail chopped off, so that it turned into a planet. . . .

Scientific eccentrics have always been with us, and can be genuinely, albeit unintentionally, amusing. They are of many varieties, ranging from the Flat Earthers to the Flying Saucer devotees, the astrologers, the hollow-globers and so on.[20] Dr Velikovsky is an almost perfect example of the pseudo-scientist. Amazingly, his book was taken seriously by some critics, possibly because it had been published by a reputable firm (Macmillan). John J. O'Neill, science editor of the *New York Herald Tribune*, described it as 'a magnificent piece of scholarly research',[21] while Ted Thackray, editor of the *New York Compass*, went so far as to compare Velikovsky with Galileo, Newton, Kepler and Einstein.

Subsequently, Velikovsky became something of a cult figure in the United States, and he produced other books following up his original theories, each of which was weirder than the last. Eventually there was a meeting at which Velikovsky was confronted by various astronomers, and the proceedings were published in 1977.[22] All the Velikovskian hypotheses were discussed—such as his claim that during their various encounters in Biblical times Venus, the Earth and Mars exchanged atmospheres, and that these encounters heated Venus to incandescence, so that it is now cooling down quickly enough for the drop in temperature to be measured even over periods of a few years.

The trouble about all this, of course, is that Velikovsky's ignorance of astronomy (and, indeed, science in general) was so complete that there is no common ground upon which rational arguments can be based. A comet is of extremely low mass by planetary standards, and the idea that a comet can change into a planet, or vice versa, is absurd. Neither is there any need to dwell upon the mathematical impossibility of a body such as Venus having its orbit changed from an erratic ellipse into an almost perfect circle, or the equal impossibility of a planet being shot out from some Jovian volcano. The whole episode is interesting psychologically, but it is difficult to see how Velikovsky's theories could ever have been taken seriously. No more need be said!

From Venus, the Earth would naturally be an outer or superior planet, and if it could be seen from the Cytherean surface it would be brilliant—though in fact the cloudy atmosphere of Venus would prevent any celestial body, even the Sun, from being observed. From Mars, both Earth and Venus would be inferior planets, showing phases, but the Earth would be much the brighter of the two, even though it is less reflective (its albedo is only about 40 per cent). From Jupiter and the other giant planets, Venus would remain so close to the Sun in the sky that it would be difficult to see at all. The best view of it would be obtained from Mercury; Venus would be a superior planet, and a splendid object in the night sky.

With this brief account of the movements and phases of Venus, let us now turn to telescopic observations of the planet itself.

4 Early Telescopic Views of Venus

Glorious as it is to the naked eye, Venus is much less spectacular when seen through the telescope. Saturn shows its rings, Jupiter its belts and satellites, Mars its dark areas and its polar caps, but Venus appears almost featureless. Very often nothing can be seen apart from the characteristic phase, and the few surface markings visible from time to time are always elusive and ill-defined. The reason is not far to seek. We are not looking at a solid surface; all we are seeing is the upper part of a cloudy layer which never clears away. The description given as long ago as 1850 by Sir John Herschel can hardly be bettered:[1]

> 'We see clearly that its surface is not mottled over with permanent spots like the Moon; we notice in it neither mountains nor shadows, but a uniform brightness, in which sometimes we may indeed fancy, or perhaps more than fancy, brighter or obscurer portions, but can seldom or never rest fully satisfied of the fact.'

Visual observations alone can tell us little about the true nature of Venus. We must turn to spectroscopic work, and more recently, of course, to radar and space-probe results. None of these techniques had become really usable before our own century; in 1900 the spectroscope had indeed been developed to some extent, but radar was unknown, and the very idea of sending a rocket to Venus (or anywhere else) was greeted with scorn. First, then, let us sum up the results of observations made before 1900.

Galileo was the first really serious telescopic observer. From the winter of 1609–10 he ranged round the entire sky, and, as we have seen, his detection of the phases of Venus was a powerful argument in favour of the Copernican theory rather than the Ptolemaic. But even his most powerful telescope magnified a mere 30 times, and in comparison with modern instruments it gave poor definition, so that Galileo could hardly hope to see any details on Venus. Neither were his immediate successors more fortunate. Probably the best observer of the mid-seventeenth century was Christiaan Huygens, of pendulum clock fame, who was certainly the first to record detail on Mars; his drawing of the famous marking known today as the Syrtis Major was amazingly accurate in view of the small-aperture, long-focus refractor that he had to use.[2] The Syrtis Major was recorded in 1659. Yet even Huygens failed to make out anything on the brilliant disk of Venus, and this makes us highly suspicious of the positive results claimed by observers of lesser skill.

In 1645 Francesco Fontana, a Neopolitan lawyer and amateur astronomer, recorded 'a dark patch almost in the centre of the disk of Venus'.[3] Fontana's telescope was home-made, and probably not much better than Galileo's. Earlier, in 1636 and 1638, he had made similar sketches of Mars, showing a circular disk with a ring inside it and a dark patch in the centre.[2] There is no doubt that these effects, both with Venus and with Mars, were purely optical. Fontana was followed in 1665 by another Italian amateur, Burattini, who saw the same kind of appearance—again purely optical.[4]

Next, in 1667, came Giovanni Domenico Cassini, who made his observations from Bologna. He recorded various bright and dusky patches;[5] his first sketch was made on 14 October 1666, at 17h 45m, and from his observations made during this period he produced the first estimated rotation period—23h 21m, only a little shorter than that of the Earth. Subsequently Cassini left Italy to become the first Director of the newly founded Paris Observatory, and in the less transparent skies of France he was unable to recover the markings on Venus. His son, J. J. Cassini, who followed him as Director, was almost equally unsuccessful, though he did try to confirm the rotation period.[6,7]

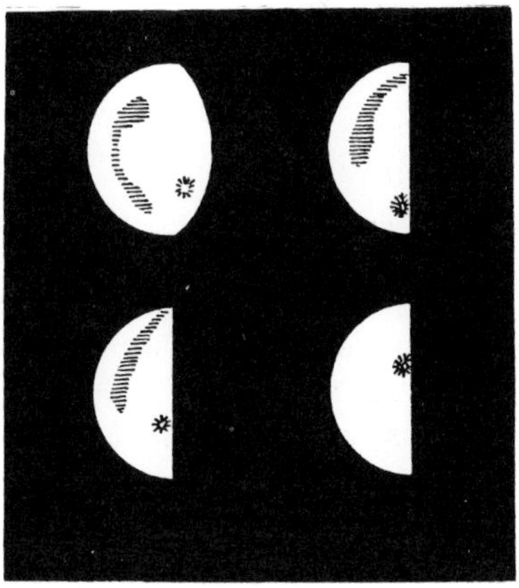

Fig. 5
Four drawings of Venus,
by G. D. Cassini.

The elder Cassini had several important discoveries to his credit, such as the main division in Saturn's ring system; he also found four of Saturn's satellites (Iapetus, Rhea, Dione and Tethys). He was unquestionably a good observer, but it is not easy to understand why he so completely failed to recover the shadings on Venus after 1667—if he had ever seen them in reality. He was much more successful in determining the rotation period of Mars, giving a value of 24h 40m, which is less than three minutes too long. It may be unfair to dismiss his Cytherean shadings as spurious, but Venus is a particularly difficult object to study with a small refractor which is bound to give a great deal of false colour, and the best we can say is that his observations of markings on Venus are of doubtful validity.

The next claims were made from 1726 by Francesco Bianchini of Rome. He studied Venus with a 6·4-cm refractor with a focal length of 20 metres, using a magnification of about 100. Bianchini's results were frankly startling.[8] They included a map showing 'oceans' and 'continents', as well as various other features. He also gave a rotation period of 24d 8h.

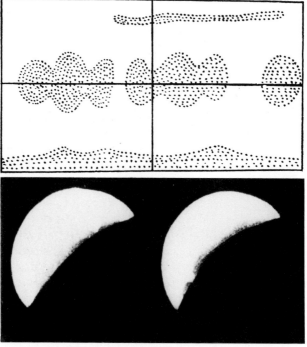

2   Observations of Venus by Bianchini. *Above*: map of the surface. *Below*: drawings of the planet. From *Hesperi et Phosphori Nova Phænomena*, Rome 1727.

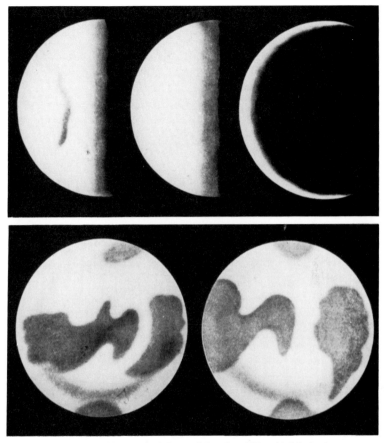

3   Drawings of Venus, by Schröter (*above*) and Bianchini (*below*).

It is very unlikely that Bianchini's sketches show any genuine markings on Venus. Again we have to dismiss them as optical effects. Moreover, his efforts to draw a map are of no scientific value except from an historical point of view.

Apart from Galileo's discovery of the phases, the first really useful telescopic result was due to the Russian astronomer M. V. Lomonosov, in 1761. In that year Venus passed in transit across the face of the Sun, and Lomonosov found that the outline was slightly blurred and hazy. From this, he stated that Venus 'is surrounded by a considerable atmosphere equal to, if not greater than, that which envelops our earthly sphere'.[9] Otto Struve has suggested that the haziness may not really have been due to the Cytherean atmosphere,[10] but in any case Lomonosov's conclusion was quite correct, and the credit for the discovery of the atmosphere of Venus must go to him, a point upon which modern Soviet astronomers are rightly insistent.[11] Rather surprisingly, Lomonosov's announcement attracted little attention at the time, and his great French contemporary Lalande was inclined to believe that Venus had no atmosphere comparable with our own,[12] but the existence of an atmosphere was virtually proved later in the eighteenth century by Johann Hieronymus Schröter.

Schröter has an honoured place in the history of astronomy. He was an amateur who set up his observatory at Lilienthal, near Bremen (where he was Chief Magistrate), and equipped it with the best telescopes he could obtain. He observed the Moon and planets systematically from 1778 until 1814, when his observatory was destroyed by the invading French army and his brass-tubed telescopes plundered by the soldiers—who mistook them for gold. Schröter was the first really great observer of the Moon, and it may be said that he laid the foundations of selenography. He was equally energetic in observing the features on Mars, though admittedly he misinterpreted them, and he was deeply involved in the hunt for the 'missing planet' between the orbits of Mars and Jupiter; indeed, the third asteroid, Juno, was found by his assistant Karl Harding, from Lilienthal. So far as Venus is concerned, his main results were contained in his book published in 1796,[13] and are of great value.

In modern times it has become almost the fashion to decry Schröter, and to claim that his work was of uniformly poor quality. Thus G. de Vaucouleurs states that 'unfortunately hardly any of his conclusions has survived the test of later progress ... thus astronomy has derived only little benefit from his industry'.[14] Yet a close examination of his surviving books and notes seems to show that this criticism is unjustified.[15] Schröter was admittedly a clumsy draughtsman, but he seldom made a serious error; his largest telescope—a 48-cm reflector by Schräder of Kiel—may have been mediocre, but two of his other reflectors were made by William Herschel, and were undoubtedly good.

Schröter did not find Venus an easy object. Between 1779 and 1788 he failed to detect any markings at all, but on 28 February 1788 he 'perceived the ordinarily uniform brightness of the planet's disk to be marbled by a filmy streak'.[13] Subsequently he saw other markings, but all were diffuse and ill-defined, so that he came to the correct conclusion that they were atmospheric in nature.

There were other confirmatory observations, as Schröter noted.[16] The light of the disk falls away perceptibly towards the terminator, which indicates absorption in an atmosphere; also, the horns of the crescent Venus are often seen to be prolonged beyond the semi-circle, which is never the case with a planet devoid of atmosphere. Schröter's reasoning was, in fact, quite sound. The fading of the light towards the terminator is almost always very obvious, and the extension of the horns can be so marked that it may occasionally stretch right round the dark hemisphere, giving the appearance of a ring. Many later observers have seen this annular

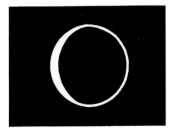

Fig. 6   Extension of the cusps of Venus.

appearance; for instance Guthrie in 1842,[17] Gibbs in 1866,[18] Lyman in 1866 and 1874,[19] and many nineteenth-century observers.[20] It was also confirmed photographically by E. C. Slipher and J. B. Edison at the Lowell Observatory.

Since the shadings on Venus are so elusive and difficult to define, measuring the rotation period of the planet is an almost impossible task by visual observation alone, but in Schröter's time no other method was available, and he did his best. He gave a value of 23h 21m 19s in 1789,[21] and modified this to 23h 21m 7s·977 in 1811.[22] Giving a value to a thousandth of a second seems rather curious, but it is only fair to add that Schröter himself was well aware of the problem, as he commented in a paper published in 1792:[23]

'The circumstance also that there are seen on this planet none of the flat spherical forms as are conspicuous on Jupiter and Saturn, none of the strips or longitudinal spots parallel to the equator which are seen on these planets . . . and which point out a certain stretch of atmosphere, give room to infer, that the globe of Venus . . . performs its rotation round its axis in a much longer space of time than these planets . . . and this is actually confirmed by my observations of the distinct part of Venus.'

Schröter also referred to rather similar comments made earlier by the French observers De Goimpy[24] and Mairan.[25] He believed the axial inclination to be about 15° to the perpendicular, as against over 23° for the Earth.[13]

Yet another confirmation of the existence of an atmosphere came from observations of the phases. In particular, there is the discrepancy between theoretical and observed dichotomy. At eastern elongations, when Venus is an evening star and therefore waning, dichotomy always occurs before the predicted time, while at western elongations, when Venus is waxing in the morning sky, dichotomy is late. Not surprisingly, it was Schröter who first called attention to this, when, in 1793, he found that the theoretical and observed half-phase differed by as much as eight days.[26] It was confirmed in the nineteenth century by Beer and Mädler, the great lunar observers, who found the average discrepancy to be six days—dichotomy being early for eastern elongations, late for western[27]—and by F. di Vico from Italy.[28]

In 1957 one of the present writers (Moore) discussed this phenomenon, and suggested that it should be called 'Schröter's Effect'. The suggestion was well received, and by now the term has become generally accepted. The effect is undoubtedly real, and it is now known that observed and theoretical phase are generally rather different, so that Schröter's effect is not confined to the time near dichotomy. More will be said about this in Chapter 5. Meanwhile, we can safely reject the suggestion by Beer and Mädler that it is due to shadows cast by high mountains on to the surface.[27] There is no chance that the theoretical predictions are wrong, so that we are certainly dealing with an effect produced by the Cytherean atmosphere.

We now come to a very interesting period in the history of the observation of Venus—the verbal battle between Schröter and William Herschel concerning the existence or non-existence of high mountains on the planet.[29]

Herschel was, of course, one of the greatest observers of all time—perhaps the greatest of all. He began his astronomical work in 1772, and in 1781 became world-famous for his discovery of the planet Uranus. It is true that he was not looking for a new planet, and did not recognize its true nature even when he saw it (he mistook it for a comet). It is also true that his main work was in connection with the stars; he was the first to give a reasonably good picture of the shape of the Galaxy, he proved the existence of physically-associated pairs of stars (binary systems), and he discovered large numbers of star-clusters and nebulæ. He was by far the best telescope-maker of his time, and his greatest reflector, completed in 1789, had an aperture of 125 centimetres and a focal length of 12 metres, though admittedly most of his best work was undertaken with smaller telescopes. His observations of the planets were more or less incidental, and confined mainly to the earlier part of his career, but he did look at Venus; a typical drawing is given here, with Herschel's own description of it:[30]

Fig. 7
Venus, as drawn by
William Herschel
on 19 June 1780.

'June 19th, 1780. There is on Venus a bluish, darkish spot *adc*; and another, which is rather bright, *ceb*; they meet at an angle, the place of which is about ⅓ the diameter of Venus from the cusp, *a*. June 21, 23, 24, 25, 26, 28, 29, 30 and July 3: Continued observations were made on these and other faint spots. . . . The instrument used was a 20-foot Newtonian reflector, furnished with no less than 5 different object specula, some of which were in the highest perfection of figure and polish; the power generally 300 and 450. But the result of them would not give me the time of rotation of Venus. For the spots assumed often the appearances of optical deceptions, such as might arise from prismatic affections; and I was unwilling to lay any stress upon the motion of spots, that either were extremely faint and changeable, or whose situation could not be precisely ascertained. However, that Venus has a motion on an axis cannot be doubted, from these observations; and that she has an atmosphere is evident, from the changes I took notice of, which surely cannot be on the solid body of the planet.'

Schröter had two telescopes which he had bought from Herschel (incidentally, of all the telescopes which Herschel made, it seems that only those which he retained, and the two which he sold to Schröter, were ever used for serious research!). Generally, the two men were on excellent terms. Their one brush came with regard to the 'mountains on Venus' which Schröter believed he had seen. The story is worth recalling in some detail here.

As we have noted, Schröter had to wait for a long time before seeing any detail on Venus. In his own words: 'I perceived neither spots, nor any other remarkable appearance, except the unusually quick decrease of light toward the boundary of illumination, which itself was not sharply defined.[31] Then, on 28 December 1789, using magnifications of from 161 to 370 on his 7-foot focus Herschel telescope, he saw that the southern cusp was blunted, while beyond it there was a small luminous speck. He saw it again on 31 January 1790,[32] and again, three times, in December 1791. By now he was fully convinced that it was an 'enlightened mountain' of very considerable height, catching the rays of the Sun—as does actually happen with the Moon, when peaks near sunrise or sunset on the lunar surface can often be seen apparently well clear of the terminator.

Fig. 8   Schröter's 'Enlightened Mountain'.

In 1790 he made regular attempts to observe the 'mountain'. What he found was that the southern cusp was generally blunter than the northern, and less regular—and this is not unusual (for instance, Moore saw it very strikingly on many occasions during April and May 1980). Schröter's account of his observations of March 1790 seems worth quoting at some length:[33]

'On the 9th of March, 1790, immediately after sunset, and till 6h 45m, I saw Venus with a 7-foot reflector, magnifying 74, 95 and 161 times, very distinctly, and uncommonly splendid. The southern cusp did not appear precisely of its usual circular form, but rather inflected in the shape of a hook beyond the luminous semicircle into the dark hemisphere of the planet.

'On the following evening, the air being as calm and serene as the preceding one, I observed the planet from 6h to 6h 40m, ... The southern cusp had its luminous prolongation, but not quite so distinct as the preceding night; but what was more remarkable, each cusp, but chiefly the northern one, had now most evidently a faint tapering prolongation, of a bluish grey cast, which, gradually fading, extended along the dark hemisphere, so that the luminous part of the limb was considerably more than a semicircle.

'On the next night ... I found Venus before sunset, with a power magnifying 95 times. At 6 o'clock, I saw distinctly the southern point terminating in a luminous streak; which now, as on the evening of the 9th, was longer and narrower than the bright termination of the northern cusp.

'The very next, or the fourth evening, gave me a favourable opportunity for this purpose. At 6 o'clock, the atmosphere being uncommonly clear, I looked at Venus with the 7-foot reflector, magnifying 95 and 74 times. It appeared very distinct, and I ascertained, beyond the possibility of doubt, that the southern cusp projected somewhat into the dark hemisphere, and that from the point of the northern one, the very faint narrow streak of pale bluish light, intermittent in intensity on account of its faintness, but yet permanent as to duration, extended several degrees along the limb of the dark hemisphere of the planet.'

Schröter then went on to interpret his observations:

'When the southern cusp extended, not in the true spherical curve of the limb of the planet, but in a somewhat hooked direction, into the dark hemisphere, the pale bluish ash-coloured streak appeared only at the point of the northern cusp, from whence it proceeded, in a true spherical curve, along the dark limb of the planet. On the 10th of March, on the other hand, when the southern cusp did not penetrate so far into the dark hemisphere, the pale streak was perceived at both points, though somewhat more sensibly at the northern than the southern; and such also were the appearances after inferior conjunction. These appearances will be thus explained by the effects of a twilight. The bright prolongation of the southern cusp, as it was seen on the 10th and 12th of March, must be ascribed to the solar light illuminating a high ridge of mountains situated at this region. . . .

'Considering the immense height of the mountains, and the great inequalities of the surface of Venus, it is natural to suppose that, at the times of its greatest elongations, one cusp frequently appears pointed and the other blunt; owing to the shadow of some mountain darkening the extremity of the latter, the same appearance may often take place in the falcated phase of the planet. But the cusp whose extremity is covered by a shadow, will, in this case, so far from appearing blunt, always exhibit a pointed appearance. . . . The shadows of mountains will, no doubt, at times occasion an uneven, ragged appearance, but cannot materially affect the very faint light of the whole.

'. . . Though we cannot suppose a smaller, but rather a greater force of gravity on the surface of Venus than on our globe, nature seems, however, to have raised on the former such great inequalities, and mountains of such enormous height, as to exceed four, five and even six times the perpendicular elevation of Cimboraco, the highest of our mountains.'

His estimate for the height of the Cytherean atmosphere was between four and five kilometres.

At this point Herschel entered the discussion—and with a sarcasm and vehemence quite uncharacteristic of him. He had been making observations of Venus ever since 1777, and had never seen anything apart from very vague shadings on the disk. In his paper criticizing Schröter,[34] he asked 'by what accident I came to overlook mountains in this planet which are said to be of enormous height'. He attributed the effects seen by Schröter along the terminator to 'a deception arising from undulations in the air', and even questioned the condition of Schröter's telescope, which Herschel himself had made. 'Probably the mirror, which was a very excellent one, was by that time considerably tarnished, and had lost much of the light necessary to shew the extent of the cusps in their full brilliancy.' He stated that 'As to the mountains in Venus, I may venture to say that no eye, which is not considerably better than mine, or assisted by much better instruments, will ever get a sight of them.'

This was certainly uncompromising. Fortunately Schröter did not rise to the bait, and his reply[35] was calm and courteous; he alluded to 'the friendly sentiments which the author has always hitherto expressed toward me, and which I hold extremely precious'. Very reasonably, he pointed out that Herschel's main attention was devoted more to stellar observations than planetary ones:

> 'I should indeed be surprised that the celebrated author had not, in all the time since 1777, perceived any inequality in the boundary of light, or other appearance of that kind, tending to confirm the existence of very high mountains according to the old observations, were it not that his bold spirit of investigation has been chiefly employed in making much more extensive discoveries in the far distant regions of the heavens, where he has gathered unfading laurels. In fact, the observations which he has communicated from his journals are *much too few* to prove a negative against old and recent astronomers.'

Finally, he pointed out that he had never actually seen mountains on Venus, but 'had only *deduced* their existence and height from the observed appearances. It is even impossible to see them, according to what I have expressly asserted in my paper on the Twilight of Venus.'

Schröter's attitude effectively de-fused a potentially explosive situation. So far as he and Herschel were concerned, the matter ended there; neither is there any hint that the two were subsequently upon anything but the best of terms.

Interpretations notwithstanding, other observers confirmed much of what Schröter had reported. Schröter had himself referred[31] to an early comment by P. La Hire that when near inferior conjunction the terminator of Venus was serrated to a greater extent than that of the Moon—though once again it seems that this was an optical effect. The bluntness of the southern cusp was often recorded—for instance by James Breen in 1853–4,[36] using the Northumberland refractor, the Cambridge telescope which will always be remembered as the instrument with which James Challis did *not* discover the planet Neptune even though all the information had been put into his hands! Terminator indentations or dips were also seen. For example, on 14 May 1892 F. Porro using the 28-cm Merz refractor at Turin, saw the terminator bulging out in the centre, with depressions to either side[37]—confirmed in the following month by W. Alexander in Britain.[38] In 1873 and 1881 a very skilled observer, W. F. Denning of Bristol, saw a dusky indentation not far from the northern cusp, 'extremely small and looks like a crater, though I could not be certain of this'.[39] Yet Denning was no supporter of Schröter's enlightened mountain, and in 1891 wrote:[40]

'There is strong negative evidence among modern observations as to the existence of abnormal features; so that the presence of very elevated mountains must be regarded as extremely doubtful, if, indeed, the theory has not to be entirely abandoned. The detached point at the southern horn shown in Schröter's telescope was probably a false appearance due to atmospheric disturbances or instrumental defects.'

However, the detached point of light away from the south cusp was reported again in 1876 by Van Ertborn of Antwerp, by Lohse and Wigglesworth with a 39-cm refractor on 2 January 1886,[41] and even, on occasions, by Denning himself. On 30 September 1876 C. V. Zenger, in Prague, described 'very high reflecting points on the surface of Venus (perhaps snow-covered peaks) being visible, as on the Moon, before the sunlight reaches the lower parts of the surrounding surface'.[42] On 17 April 1873 R. Langdon, an English amateur, reported 'two exceedingly bright spots on the crescent— one close to the terminator, toward the eastern horn, and the other in the centre of the crescent ... appeared like two drops of dew'.[43] And on 29 January 1870 Henry Pratt, from Brighton, saw 'a tooth of light near the south cusp ... evidently a spot on the terminator higher than the adjacent regions'.[44]

In fact, during the latter part of the nineteenth century Schröter's mountain theory was still very much alive. One strong supporter of it was Étienne Trouvelot, a French astronomer who spent some time at Harvard in the United States before returning to France in 1882 to work at the Observatory of Meudon. His observations of Venus were made in two series: from 1875 to 1882 at Harvard, and from 1882 to 1892 at Meudon. He wrote:[45]

[On 2 February 1878] 'The polar patches are distinctly visible, the southern one being the more brilliant. Their surface is irregular and seems like a confused mass of luminous points, separated by comparatively sombre intervening spaces. This surface is undoubtedly very broken, and resembles that of a mountainous district studded with numerous peaks. ... The polar spots seem to be bristling with peaks and needles. This is especially the case with regard to the southern spots, which seem to be entirely formed of brilliant points. On the north polar spot is a luminous peak which seems to project outside the limb; or if it is not so, the parts surrounding this peak must be dark enough to be confused with the background of the sky. But I am fairly certain that this spot projects outside the limb.'

Moreover, the same theory—of mountain-tops catching the light of the Sun—was still supported as recently as 1947 by McEwen.[46] It is rather ironic that today, when we have our radar maps of Venus, it has been shown that the surface is indeed mountainous—but the Cytherean heights bear no relation to the phenomena reported by Schröter and his supporters.

There was also the theory of a vast elevated plateau near the southern cusp, which is linked with observations of the Cytherean cusp-caps.

It is undeniable that the cusps of Venus often show bright caps or hoods. They were seen by Franz von Paula Gruithuisen in 1813 and subsequently,[47] and they can be striking. Gruithuisen believed them to be permanent, and went so far as to compare them with the well-known polar caps of Mars. But were they truly polar? At that time (and in fact until quite recently) the axial inclination of Venus was unknown, and there was no proof that the cusps marked the Cytherean poles. Yet the caps were there; almost all serious observers of Venus saw them—for instance Vogel and Lohse in 1871 from Bothkamp Observatory,[48] and Schiaparelli from 1877 to 1892.[49]

To anticipate Chapter 5, it may be added that one of the present writers (Moore) has been studying the caps ever since 1934, and well before the launching of Mariner 2 was maintaining that they could only be polar, due to the circulation of the Cytherean atmosphere in high latitudes—a conclusion which has turned out to be correct. This is, of course, very different from assuming them to be snow-covered, elevated plateaux, as was believed by Trouvelot. During the transit of Venus across the face of the Sun in 1882, F. Arago and B. de la Grye took photographs[50] indicating a bulge in the disk near the southern cap—interpreted by Trouvelot as being a plateau with an elevation of over 100 metres, while de la Grye attributed it either to a vast accumulation of ice and snow or else to a huge cloud-mass. Another feature was a band along the edge of either cap, described by Percival Lowell as 'a sort of collar'.[51] It was also seen by other observers, such as Niesten and Stuyvaert from Brussels.[52] This also can be quite prominent, though it was (and still is) widely attributed to the effects of contrast.

Meanwhile, continued attempts were being made to measure the length of the rotation period of Venus—purely by observations of the surface markings, vague and unsatisfactory though they were. Generally speaking, the estimated periods were of the order of 24 hours, and some of them were taken to an absurd degree of accuracy. At Lussinpiccolo, in 1895, Leo Brenner gave a value of 23h 57m 36s·2396,[53] amending it in the following year to 23h 57m 36s·27728.[54] One cannot help feeling that this is rather like estimating the age of the Earth to the nearest minute. (As a matter of fact Brenner—an assumed name—was very much of an astronomical charlatan, and the accuracy of any of his published observations is, to put it mildly, dubious.)

Then, however, G. V. Schiaparelli, the Italian astronomer who will always be remembered for his reports on the canals of Mars, produced an astronomical bombshell in the form of a rotation period for Venus of 224d 16h 48m,[55] which is precisely the time taken for Venus to complete one orbit round the Sun. In other words, Schiaparelli came to the conclusion that Venus keeps the same hemisphere turned permanently sunward.

Schiaparelli was undoubtedly a skilful observer, even though he was so completely wrong in drawing hard, sharp, even double linear features on Mars. His main observations of Venus were carried out from Milan in 1877–8 and in 1892; as with Mercury, he observed mainly in daylight, so that Venus was high in the sky. From his studies of the southern cusp-cap, he derived what he believed to be a definite rotation period. In fact there was nothing particularly surprising about it. The Moon behaves in just such a way relative to the Earth, and so do the main satellites of the other planets with respect to their primaries. Tidal friction over the ages is responsible, and the idea of Venus having a similarly captured or synchronous rotation was quite reasonable. Schiaparelli also studied the delicate surface markings on Mercury, and came to a similar conclusion—a rotation period equal to the revolution period (in the case of Mercury, 88 Earth-days). His opinions carried a great deal of weight, and for Mercury the synchronous rotation was still accepted until radar work in the early 1960s showed it to be wrong (the true value is 58·56 Earth-days). Venus was more of a problem; some observers agreed with Schiaparelli, while others did not. It is worth noting, however, that as late as 1955 the synchronous rotation period was still accepted by A. Dollfus, one of the leading modern observers of the planets.[56]

It is interesting to look back at the various rotation periods proposed for Venus before the problem was finally cleared up. Estimates ranged between 24 hours and 225 days. The one thing that was *not* suspected was a rotation period longer than the Cytherean year—and yet this proved to be the truth.

Let us now come to the spoke-like markings of Venus, sometimes nicknamed the Cytherean canals. They do not exist, and they were never regarded as artificial; but the story is worth examining, and inevitably it is to some extent linked with the famous (or infamous) canals of Mars.

The Martian network of canals was first reported in detail by Schiaparelli in 1877. They were described as being unnaturally regular, forming a planet-wide réseau, and subject to doubling or 'gemination'. So far as their origin was concerned, Schiaparelli kept an open mind; his own name for them was 'channels' (*canali*), though inevitably it was translated as 'canals', and Schiaparelli was careful not to dismiss the idea of their being artificial. He certainly believed them to be channels through which water flowed equatorward from the shrinking polar caps during spring and early summer on Mars. They were confirmed in 1886 by Perrotin and Thollon, at the Observatory of Nice, and from then onwards they became thoroughly fashionable. Not until the Mariner results of the 1960s were they finally banished to the realm of myth.

Schiaparelli was followed by Percival Lowell, who founded his observatory at Flagstaff in Arizona principally to observe Mars. Between 1894 and his death in 1916, Lowell and his colleagues recorded hundreds of Martian canals, and attributed them to channels built by intelligent engineers in an attempt to defeat the growing water shortage on Mars by means of a vast irrigation scheme.

Lowell's observations, as well as his conclusions, were hotly challenged. Other observers, using telescopes as powerful as Lowell's 61-cm refractor, either failed to see the canals at all, or else recorded them as hazy, irregular streaks. Moreover, Lowell showed straight lines not only on Mars, but upon other bodies such as Mercury, the satellites of Jupiter—and Venus.

Vague streaks on Venus had been recorded by Perrotin at Nice,[57] but Lowell's chart of the planet, published in 1897 with an accompanying description, was much more definite.[58] He even named the features which he believed he had seen. From a dark patch, Eros—a sort of focal centre—he drew well-defined dark strips which he named Adonis Regio, Æneas Regio and so on. He was quite convinced that the markings were unchanging, and that the captured or synchronous rotation period was correct, so that the same hemisphere was permanently sunlit. His description of the markings runs as follows:

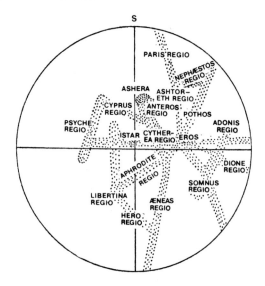

Fig. 9   Lowell's map of Venus.

'The markings themselves are long and narrow; but unlike the finer markings on Mars, they have the appearance of being natural, not artificial. They are not only permanent, but permanently visible whenever our own atmospheric conditions are not so poor as to obliterate all detail on the disk. They are thus evidently not cloud-hidden at any time. . . . There is no distinctive colour in any part of the planet other than its general brilliant straw-coloured hue. The markings, which are of a straw-coloured grey, bear the look of being ground or rock, and it is presumable from this that we see simply barren rock or sand weathered by æons of exposure to the Sun. The markings are perfectly distinct and unmistakable, and conclusive as to the planet's period of rotation. There is no certain evidence of any polar caps.'

Lowell thus rejected the whole idea of a dense, concealing atmosphere round Venus, and he maintained this view in his many papers on the subject.[59] Others did not agree. Even Brenner believed the streaks to be optical effects, vanishing under conditions of good seeing.[60] In 1897 Antoniadi, who always gave his views very frankly, wrote a paper[61] to which he added a somewhat caustic postscript, as follows:

'I refrain from discussing the aphroditographic* work of most of our contemporaries, who, forgetting that Venus is decently clad in a dense atmospheric mantle, cover what they call the "surface" of the unfortunate planet with the fashionable canal network, dividing it into clumsy melon slices having their radiant now on the cusps and then on the visual ray. Such discussion would lead to some curious conclusions in which we should not be long in finding that Venus does not constantly present the same face to the Sun—but to the Earth.'

It is often said that men are remembered not for their achievements, but for their mistakes. Lowell is a case in point. He was a good organizer, a good mathematician and a brilliant writer and speaker. Yet it must be said, with regret, that he was not a reliable observer, and he unquestionably tended to see hard, straight features which did not actually exist. There was nothing the matter with his telescope; the 61-cm refractor is still in perfect order (Moore has used it extensively, and has made many drawings of Mars with it, though without seeing any trace of canals). Antoniadi, however, used an even larger telescope, the Meudon 83-cm refractor, and he never felt it necessary to stop down the aperture, as Lowell did at Flagstaff. Another sceptic was E. E. Barnard, renowned for his keen eyesight. Using the Lock Observatory 91-cm refractor, he wrote:

'Venus has been examind on a number of occasions with the 36-inch, when the planet was beautifully defined. . . . Nothing was seen of the singular system of dark narrow lines shown in recent years by observers to cover the surface of the planet. Every effort was made to show them, by reducing the aperture and by the use of solar screens and various magnifying powers. They were also looked for with the 4-inch finder. Previous attempts with the 12-inch here also failed.'

A. E. Douglass, who worked at the Lowell Observatory and who also had recorded the linear features, was quick to take up the

---

*The hideous adjective 'aphroditographic' is perfectly correct. One can only hope that it will never come into general use.

cudgels, and in 1898 he published a strongly-worded reply.[62] His comments are worth quoting:

'The reading public has been recently addressed on the subject of the markings on Venus in various attempts to show that the discoveries made at this observatory are unworthy of credit. No matter how futile such criticism must prove to be in the long run, some persons will be influenced by it if from time to time we do not make some rejoinder, or give out some statement which will show our continued activity in this line of work, our undiminished confidence in the results obtained, and our answering attitude toward adverse opinion.

'In the last six years many thousands of hours have been spent by us at telescopes of 13, 18 and 24 inches aperture and their smaller finders, when the seeing was sufficiently good for profitable work on the finest known planetary detail. . . . Under proper conditions of air and atmosphere the markings on Venus are absolutely certain. Under proper conditions they are to me about as easy or difficult to see as the irregularities in the terminator of the Moon when it is near first quarter, viewed by the naked eye. I have on a few occasions seen a large projection perfectly distinct. At the best seeing the markings are visible at the first glance.

'To say that no markings save M. Antoniadi's symmetrical shadings of atmospheric contrast exist, or that the detail seen here is due to pressure on our objective, or to defective densities in the eyepiece, or to our own eyes, or to the imaginings of our brains; or, most ridiculous of all, to our looking all day at some map and then seeing it on the planet, is to offer suggestions too absurd to be taken seriously.'

Douglass then went on to catalogue the reasons why, in his view, the markings had not been seen by other observers. He discussed atmospheric conditions, to which he believed insufficient attention was paid, and also the fact that there were not many observers who had studied Venus as constantly and over as long a period as the Flagstaff team. But he also made a very peculiar statement:

'The second reason why some observers have not seen them has been the fault of using too large an aperture. . . . I decided long since that in planetary work the greatest efficiency is obtained with the smallest aperture which supplies the required illumination. There is a limit to this, however. An inch and a half lens shows the markings on Venus nicely, but they are not so well defined as in a lens of 3 inches, which in our atmosphere is a very satisfactory size to use. When the seeing is very bad an aperture of less than 3 inches will become necessary.'

The idea that a small, portable telescope can show as much on Venus (or anything else) as a great refractor or reflector is, frankly, ridiculous. Moreover, using a small aperture on an object as brilliant as Venus—even when observed in full daylight—is an open invitation to produce optical effects. No observers, except those at Flagstaff, ever recorded the spoke-like features on Venus with large telescopes. Yet, surprisingly, the arguments continued well into our own century, and it may be as well to carry the story to its end in the present chapter.

After a long hiatus, a well-known English amateur astronomer, Robert Barker, revived the Lowell-type features in 1932. Using a 31-cm reflector, he wrote that his observations 'confirmed the dedication of Schiaparelli and Lowell that Venus' rotation synchronizes with her revolution round the Sun',[63] and in further papers[64] Barker published drawings showing some of the features on Lowell's map, such as Ashera, Hera Regio and Æneas Regio. He was supported later by R. M. Baum, present Director of the Inner Planets Section of the British Astronomical Association. Baum's drawings,[65] made with a 15·2-cm reflector and a 7·5-cm refractor, showed streaks which were broader and more diffuse than Lowell's, and Baum regarded them as atmospheric phenomena; but again we have a central dark patch from which streaks radiate outwards, and this at once indicates an optical effect rather than a genuine system of markings on Venus. An American observer, O. C. Ranck, used a 10-cm refractor to make drawings strikingly similar to Lowell's.[66] Also in the United States, J. C. Bartlett reported streaks, and claimed that 'the general picture of Venus which Lowell gave us in the 1890s is one that is easily recognized today by every close student of the planet, including those who have never heard of his Venusian work. We must conclude therefore that in the main he was factual and correct with respect to the markings on Venus.'[67]

This, of course, was in the period before professional astronomers in general began to pay much attention to the surface details on the planets; space-travel was still a dream of the future, even with unmanned probes. In 1955 and 1956 there was some animated correspondence in the columns of the *Strolling Astronomer*, the official organ of the Association of Lunar and Planetary Observers, between Moore,[68] who was openly sceptical about the whole radial system; Baum,[69] who believed in it; and Bartlett,[70] who referred to Baum's independent 'recovery' of the Lowellian spoke system, and was convinced of its reality.

Apparently the last professional astronomer to support anything in the nature of a spoke system was A. Dollfus,[71] who found that the dusky markings on Venus often showed the pattern of a radial system with its centre at the sub-solar point. He even produced a map showing what he regarded as permanent features, and another map was drawn in 1961 by K. Brasch, based on work carried out by observers of the Royal Astronomical Society of Canada.[72] But, as with Mars, the evidence of the unmanned space-probes has been conclusive. The Lowell-type markings on Venus are just as unreal as the Martian canals.

Mistakes of this kind were inevitable in the days when the only way to examine planetary surfaces was by visual or photographic observation from the Earth. And there can be no doubt that Venus presents exceptional problems; its great brilliance, its lack of well-marked details, and its nearness to the Sun make it difficult to study. Remember, too, that it is only within the last twenty years that we have found out what kind of a world it really is.

5 Venus in the Twentieth Century: Observational Results

Until well into the twentieth century, almost all our meagre knowledge of Venus was obtained by visual observation at the eye-end of a telescope. Useful photographs were taken from 1923 onwards; the first really good spectroscopic work was carried out in 1932, when carbon dioxide was identified in the Cytherean atmosphere; temperatures were measured by means of thermocouples; and the first radar contact was made in 1961. But it was not until the flight of Mariner 2, in 1962, that we had any really reliable information.

Obviously, telescopic observation is much more difficult with Venus than with most of the other planets. It is also, one has to admit, less valuable; but some interesting results have been obtained, and are well worth describing here. The general lines of research were concentrated upon the dusky shadings, the bright regions, the cusp-caps and the Schröter effect, and almost every serious observer of the planet made efforts to determine the rotation period, though the results showed a complete range of estimates from less than a day to a full Cytherean year.

It is wrong to say that nothing whatsoever can be seen on Venus by using ordinary telescopes. True, the disk may sometimes appear blank, but at other times the shadings are quite easy to detect, even though they are never easy to define. In 1953 Sir Harold Spencer Jones, then Astronomer Royal, stated that 'nothing more than faint ill-defined shadings may be seen, and these can only be seen occasionally',[1] but this is an over-statement. More often than not, some vague, elusive features will be glimpsed—provided that conditions are good, and that the telescope is of sufficient aperture.

There have been various suggestions that the shadings are not real, but are due solely to optical effects. Experiments with 'artificial planets' have been somewhat contradictory. The first experiments of this kind seem to have been carried out in 1897 by the Swiss astronomer W. Villiger, of the Zeiss optical works at

Jena.[2] Villiger made some balls out of plaster of Paris, each about 5·6 cm across, and looked at them from a distance of 400 metres, using a 12·7-cm refractor; he duly noted dusky patches, limb bands and cusp-caps, though actually the 'artificial planets' were completely featureless. Similar results were obtained in 1952 by W. W. Spangenberg in Germany,[3] who recorded cusp-caps, collars, limb brightenings and dusky streaks when viewing his artificial planets under conditions equivalent to magnifications of from 60 to 140 on the real Venus. A third experiment was conducted in 1954 by A. P. Lenham and J. H. Ludlow.[4] Here, two transparencies of Venus, 3·2 cm in polar diameter, were made—one gibbous, backed by a filter to give it a yellowish tone. Apart from slight uniform mottling, due to the texture of the paper, the disks were blank. They were illuminated from behind, and viewed from a distance of 10 metres with a 2·5-cm telescope and a magnification of 6, giving an image of apparent size similar to that of Venus as observed with a power of 100. Twelve people—none of whom, apart from Lenham and Ludlow themselves, had any astronomical knowledge—drew the transparencies through the telescope, again showing blunted cusps, bright perimeters and central dark areas, so that the authors concluded that 'certain of the observed features of Venus may not be intrinsic in that planet'.

Moore tried the same kind of experiment, using model planets observed through a 7·6-cm refractor with different results, as no non-existent features were seen;[5,6] but unconscious prejudice is very difficult to eliminate! And though tests of this kind are not devoid of interest, it cannot be claimed that they give any really conclusive evidence one way or the other.

Yet despite Douglass' statement, quoted in Chapter 4, it seems quite definite that very small telescopes will show no genuine markings on Venus. Instruments of considerable aperture are needed, though even they have their limitations. From observations with the great 91-cm Lick refractor, E. E. Barnard wrote:[7]

'surface markings were nearly always present, but they were always very elusive, and at no time could a satisfactory drawing be secured. . . . I am confident that the faint elusive spots seen with the great telescope were real, but whether they were of a permanent nature it was impossible to tell, for the same spots were not recognized with certainty at different observations. The impression, however, was that they were not permanent.'

This description comes from a world-famous astronomer noted for his keen sight, using one of the most powerful telescopes then in existence, but it might equally well have been written by an amateur using a 15-cm reflector.

The 83-cm refractor at the Observatory of Meudon, near Paris, is particularly well suited to planetary work, and it was used extensively between 1900 and 1939 by E. M. Antoniadi, probably the best planetary observer of his time (for instance, his map of Mars has turned out to be remarkably accurate). One interesting set of observations was made in May 1928.[8] On 25 May, at 9h 50m, Antoniadi recorded a shady area near the southern part of the disk. Venus was then approaching superior conjunction, so that it was in the gibbous stage. East of the main shading lay a darker patch, with a bright area near the eastern limb. Three hours later, at 12h 45m, Antoniadi recorded the same features, sensibly unaltered; but on the following day, at 9h 10m, the main patch had vanished, leaving only the condensation and the bright area.

Visual observations of Venus with large reflectors are rather rare, but some have been made. For instance, on 28 August 1956 A. P. Lenham[9] and G. P. Kuiper made independent drawings with the 208-cm reflector at the McDonald Observatory, Texas, with a magnification of 900; no definite features were seen, despite good daylight conditions, but a general fine-scale mottling was suspected over the whole disk. Yet this is no proof that large telescopes will not show the Cytherean features, and it may be that at that moment Venus really was more or less blank. Moore, using his 32-cm reflector in England, observed it within a few hours of Lenham and Kuiper, and recorded it as featureless.

Now and again, features of unusual prominence are seen. One typical case was recorded in 1924 by W. H. Steavenson,[10] using a power of 280 on a 15·2-cm refractor:

'I at once saw a marking which was much more prominent than any I have seen before. Its most conspicuous portion took the form of a broad dusky band stretching westwards toward the limb from just south of what would be the centre of the true disk. . . . It should be visible in any telescope over 7·6 cm aperture.'

Features as conspicuous as this might be expected to yield a rotation period; but again the evidence was conflicting. In 1921 W. H. Pickering, from visual studies, announced a period of 2 days 20 hours.[11] This was supported in 1924 by Henry McEwen,[12] who devoted a lifetime to the study of the planet. McEwen, indeed, regarded the problem as definitely solved, and wrote that 'it is due to the skill and acumen of Professor Pickering that we owe the discovery of this unique rotation period ... which will be appreciated at its due value by a future generation of astronomers.'

Unfortunately the 'discovery' was not nearly so conclusive as McEwen thought. Steavenson's results were completely different. In his subsequent paper, Steavenson wrote:[13]

> 'Fortunately, the streaks and patches observed in 1924 were unusually long-lived, while their changes of form were sufficiently gradual to enable them to be identified with a reasonable amount of confidence on several (sometimes successive) nights. . . . My observations of 1924 do not tend to confirm the periods derived by Professor Pickering and Mr. McEwen. Each of my three markings happens to have been observed in almost the same position on the disk on two separate occasions, as follows: first marking February 26th and March 5th, interval 8 days; second marking March 9th and March 17th, interval 8 days; third marking March 22nd and March 30th, interval 8 days. From this it would appear that the rotation period is either 8 days or some sub-multiple of this amount.'

Further studies led Steavenson to suppose that the period actually was 8 days and not a sub-multiple, and that in the spring of 1924 'the pole was evidently turned nearly towards the Earth, and the axis lay quite near the plane of the orbit'.[14]

Pickering, McEwen and Steavenson agreed on one point: they all thought that the axial tilt of Venus must be very different from that of the Earth, more nearly resembling that of Uranus. Yet Antoniadi, from visual studies with the Meudon refractor, discounted Pickering's value of 85°, and claimed that 'the rotation axis of Venus does not form a considerable angle with the perpendicular to the plane of the orbit'.[15]

There was equal conflict about the nature of the darkish shadings. Little reliance could be placed upon the 'maps' showing allegedly permanent or at least semi-permanent features, such as that by J. Camus from Meudon in 1932, showing bright polar zones and dark and light regions round the equator,[16] or L. Andrenko from Kharkov in 1935.[17] These maps were admittedly very different from Bianchini's of more than two centuries earlier, but it must be said that charts of such a kind are of limited value at best.

Before the Space Age, there were various widely differing interpretations of the dark regions. McEwen suggested that 'Venus' atmosphere may be transparent enough to show markings ... situated (probably) in the lower and denser part which corresponds to, say, the Earth's troposphere. These may be due to volcanic smoke or dust, or to seeing dimly contrasted features which are on the surface of the planet.'[18] A. Dollfus believed that the dark, drifting, lower-lying clouds could be due to surface markings seen at times through the cloud layer.[19] H. C. Urey wrote that 'the dark areas observed by Dollfus (could be) low-lying land, awash with seas ... in this case they may be observable either as bare low flat land or as such land covered with vegetation'.[20] There were also sceptics, such as Moore, who maintained that the shadings were nothing more permanent than cloud phenomena in the upper atmosphere of Venus.[21]

4
Venus; drawing
by Patrick Moore,
39 cm reflector.

Bright areas are also seen, and there have been various reports of colours: yellows (Jarry-Desloges in 1922,[22] P. B. Molesworth in 1897,[23] and T. E. R. Phillips in 1924[23]), reds and even blues (Jarry-Desloges, Molesworth and McEwen on various occasions between 1895 and 1930).[23] Yet it seems that such reports must be treated with great reserve. With a brilliant object such as Venus, false colour is only to be expected, even if the planet is observed in full daylight.[24] We must also discount the reports of brilliant, well-defined star-like patches seen from time to time, usually by observers using small refractors. The genuine bright patches, like their dark counterparts, are of appreciable size; they are diffuse and generally rather short-lived. They may appear at any time on any part of the disk, but their vague nature means that their positions are difficult to fix accurately.

Many attempts have been made to detect movements in these bright areas. One such observation was made on 6 March 1928, by the Rev. J. A. Lees,[25] who followed a movement, in a period of two hours, of a white area from the terminator towards the limb.

A conspicuous bright area was observed by a very experienced amateur, R. L. Waterfield, on 21 August 1956, at 12h to 12h 30m, and was described by him as 'a very bright and fairly well outlined spot on the limb, just short of the north cusp. I have no doubt of its objective reality—seeing was fair, and good in flashes, when the marking stood out more definitely.... I think this is the most sharply defined feature I have ever seen on Venus.'[26] On 26 August, at 11h 45m, Waterfield reported that 'the light area was still very obvious, but it was much less brilliant, much less sharply defined, and considerably smaller. It extended over a smaller area, but was, as before, definitely short of the cusp.' On 8 September, Waterfield observed again: 'In flashes of good seeing I could see the white spot still there—and as far as I could tell, the same as when last seen on 26 August, i.e. less bright and smaller than on 21 August.'

At 6h on 26 August Moore, using his 32-cm reflector, made an observation of Venus, and saw the spot exactly as described by Waterfield (this was an independent confirmation, as Waterfield's letter was not received until later). Like all patches on Venus, this one was of limited duration, and by 20 September it had disappeared.

Occasionally, a bright patch close to the limb or terminator will cause an apparent irregularity in the outline of the planet. The bright region will appear to project, while the neighbouring darker regions give the impression of being indented. This is a pure contrast effect (it is also shown by the bright polar caps of Mars, which often seem to protrude beyond the mean disk), and, as we have seen, has led unwary observers to believe that they have discovered vast elevated Cytherean plateaux.

The cusp-caps have been closely followed, largely in attempts to derive the value of the axial inclination of Venus. In 1924 Steavenson suggested that at times the pole might be in the middle of the disk as seen from Earth, as can happen with Uranus.[27] Photographic studies by Kuiper in 1954 indicated that the equator was tilted from the orbital plane by about 32°, with a maximum uncertainty of a mere 2° either way;[28] a year later Dollfus claimed that 'the globe turns upon an axis only slightly inclined to the plane of the orbit'.[29] (Older determinations of the tilt, based purely on visual observations, are of some historical interest; 15° by Bianchini in 1727,[30] 53° by Schröter in 1796,[31] and so on.) Other estimates were made by Pickering, who gave 5° in 1920;[32] Jarry-Desloges, $45\frac{1}{2}$° in 1922;[33] Antoniadi, 60° to 80° in 1934;[34] Haas, over 75° in 1942;[35] Schwirdewahn, 0° in 1949;[36] Kutscher, 38° in 1954.[37] Ross, from his photographic work of 1927, gave a value of about 90°.[38] The suggested inclinations ranged through a complete right angle, so that at least there was plenty of variety—but at that time there was no proof that the cusp-caps were genuinely polar.

For that matter, there was no conclusive proof that the caps were anything more fundamental than contrast effects. This was the opinion of Antoniadi, in 1934, who studied them with the Meudon refractor, and claimed that the cusps appeared bright 'because of the superior glare of the limb, and because of the fact that they are bordered on two sides by the dark sky'.[39] Moore disagreed, pointing out that on this theory a cap visible at a particular phase should again be visible at the same phase at a future elongation, provided that the conditions of observation were uniform.[40] This did not seem to be the case.

It was also suggested that the caps showed definite shifts over short periods. For instance, between 6 April and 20 May 1956, W. A. Grainger found that the southern cap showed a slow but steady rotation, so that at times it was carried well away from the actual cusp.[41] The same sort of effect was noted by J. Hedley Robinson.[42] Long-continued studies by members of the Mercury and Venus Section of the British Astronomical Association indicated a general pattern of visibility for each cusp-cap with a period of about eight years.[43] Moore found that the collars were well seen only when the cusp-caps themselves were at their most prominent.[44]

Various interpretations were proposed. For instance, in 1926 Steavenson suggested that the caps might be phenomena of afternoon mist,[45] and added: 'I have observed that they do not share in the movements of other markings. The same sort of thing is occasionally to be seen on Mars. In the case of Venus, one cusp is generally more affected than the other, which suggests that the white area is not entirely an illusion. If it were, it would surely affect both cusps equally.' In 1953 a group of Russian observers proposed that the caps might be due to the relatively high luminosity of the Cytherean atmosphere,[46] due to the high reflectivity of the surface below, and that the cap was visible only when the plane through the planet perpendicular to the line of sight passed through the cap itself; they also deduced another angle for the axis of rotation—51° this time. There were, in fact, all sorts of comments and theories about the cusp-caps.[47] But it was only with the Mariner 2 findings of 1962 that it was finally shown that they are truly polar, and that the axis of rotation is inclined to the perpendicular to the orbital plane by only two or three degrees.

The Schröter effect was also closely studied. The time of actual dichotomy is not easy to define accurately, because the terminator seldom appears completely regular; generally it looks straight for several consecutive days, and even the various measuring devices which have been used do not help much. Large discrepancies were noted now and then; a fortnight in 1927 according to McEwen and Lees,[48] 12 days in 1951 according to the American observer S. C. Venter,[49] and so on. M. B. B. Heath, a very experienced and skilful observer, discussed the whole problem thoroughly in 1955,[50] and pointed out that:

> 'at the time of dichotomy the terminator is always heavily shaded owing, no doubt, to the obliquity of the solar rays there, absorption and diffusion of light in the Cytherean atmosphere, and the fact that no refracted ray can possibly reach the terminator. Consequently, the extreme edge of the terminator may be missed altogether. It is possible that the discrepancy between observed and theoretical dichotomy may be largely, or even wholly, due to these causes combined.'

Heath's own observations from 1927 to 1958 gave a general discrepancy of about 2 days, but sometimes reached 6 or 7 days, while the German astronomer W. Sandner gave an average of 4 days.[51]

In 1957 V. A. Bronshten summarized[52] the work of two energetic Russian amateurs, N. N. Michelson and V. N. Petrov, who measured a large number of drawings of Venus, and found that the phase anomalies are not limited to the period near dichotomy; this also has been confirmed on many occasions.

Results from observations made by members of the Mercury and Venus Section of the British Astronomical Association have been

informative. From 1956, efforts were made to determine the value
of the Schröter effect for each elongation. The results from 1956 to
1972 were as in the table below.

Eastern Elongations

Year	Date of Elongation	Schröter effect (days early)	Ref
1956	12 Apr	4	53
1957	18 Nov	4	54
1959	23 June	6	55
1961	29 Jan	8	56
1962	3 Sept	12	57
1964	10 Apr	7·3	58
1965	15 Nov	8·2	59
1967	21 June	5·9	60
1969	26 Jan	5·8	61
1970	1 Sept	11·1	62
1972	8 Apr	5·8	63

Western Elongations

Year	Date of Elongation	Schröter effect (days late)	Ref
1958	8 Apr	2	64
1959	11 Nov	3	65
1961	20 June	4	66
1963	23 Jan	4·5	67
1964	29 Aug	5	68
1966	6 Apr	7·2	69
1967	9 Nov	11·9	70
1969	17 June	5·5	71
1971	20 Jan	4·3	72
1972	27 Aug	4·5	73

This gives a mean of 7·1 days early for eastern elongations, 5·2 days
late for western; full details of the observations are given in the
B.A.A. Memoir.[43] Venus was excellently placed in the spring of
1980, when Moore found dichotomy to be 5 days early. Obviously

the Schröter effect is quite pronounced, presumably due to the Cytherean atmosphere.

A rather strange anomaly was discussed by Brinton and Moore in 1963.[74] During observations made soon after dichotomy, it was noticed that there was an increased difficulty in determining phase. The image should have been almost exactly a semi-circle; in fact it appeared as a segment bounded by a chord smaller than a diameter. Such an appearance could only result from viewing a sphere which was less than half illuminated, which on the face of it is absurd. However, the effect can be traced also on some photographs—an extra indication that visual observations of Venus, as well as ordinary photographs, have to be treated with some caution!

Photography, indeed, was initially of little help in unravelling the problems of Venus. The most promising method of attack was to take pictures in the light of selected wavelengths. Long waves ('red') are penetrative, while short waves ('blue') are not. Thus when Mars is photographed in red or infra-red light, the surface details are shown clearly, while in violet and ultra-violet no details can normally be seen, since these wavelengths fail to penetrate even the relatively thin Martian atmosphere.

5
Photograph of Venus;
Palomar 5·9 cm reflector.
No details are shown.

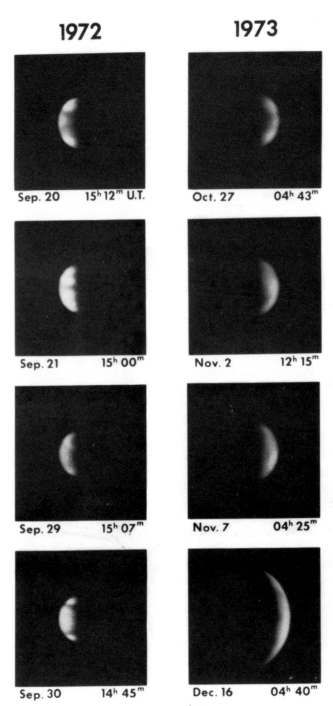

1972

Sep. 20 15^h 12^m U.T.

Sep. 21 15^h 00^m

Sep. 29 15^h 07^m

Sep. 30 14^h 45^m

1973

Oct. 27 04^h 43^m

Nov. 2 12^h 15^m

Nov. 7 04^h 25^m

Dec. 16 04^h 40^m

6 Photographs of Venus in ultra-violet light.
International Planetary Patrol Programme, Flagstaff, Arizona.

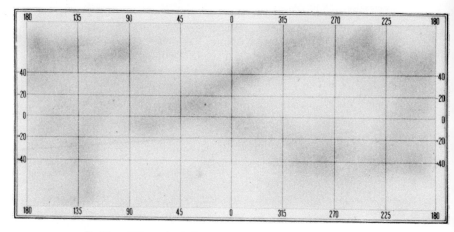

7   Map of Venus, by C. Boyer, from photographs taken at the
Pic du Midi Observatory. The Y-feature is clearly shown.

With Venus, it was reasonable to expect that infra-red
photographs might penetrate the atmosphere to some extent, but
these hopes proved to be ill-founded. Infra-red pictures showed no
more than those taken in integrated light. In 1924 W. H. Wright
experimented[75] by taking photographs with the aid of filters, and
then, in 1927, an excellent series was obtained by F. E. Ross at
Mount Wilson,[76] using the 152-cm and 254-cm reflectors. The main
point of interest was that distinct features on the disk were seen not
in red light, but in ultra-violet. Presumably, then, the best—indeed,
the only—results were to be obtained with respect to the uppermost
cloud-layer. It is most unfortunate that the original Ross plates have
been lost.[77]

Little further work of this kind was done for some years, but then
some further excellent photographic results were obtained by G. P.
Kuiper in 1950, at the McDonald Observatory in Texas;[78] N. A.
Kozyrev in the Soviet Union;[79] A. Dollfus at the Pic du Midi;[80] and
R. S. Richardson at Mount Wilson.[81] Kuiper's photographs, like
those of Ross, showed a vaguely banded appearance; so did
Richardson's. Moreover, Richardson made a valuable and at least
partly successful attempt to correlate his photographs with visual
drawings made independently at about the same time.

Dollfus, who carried out a long visual and photographic study,
came to the conclusion that 'the atmosphere of Venus contains, at
least in its upper regions, cloudy veils of appreciable density; they

are scattered, and in motion; their altitude is several kilometres. In the clearer patches between these clouds, the lower layers of the atmosphere are confusedly seen, as well as the ground, marked by patches whose distribution may be drawn.'[82] This shows, surely, the limitations of our knowledge of Venus before the era of space-probes.

Strangely enough, the best results of all at that period came from visual work by the French observers, who recorded a characteristic Y-shaped dark feature centred on the equator.[83] Boyer and Guèrin found that the Y-marking tended to persist over long periods,[84] and from it and other observations they deduced a rotation period of 4 days, in a retrograde sense. This was an outstanding result. The upper clouds really do have a 4-day period, even though the planet itself spins so much more slowly.

So much, then, for visual observations, which were inevitably limited in the amount of information they could yield. Next, let us turn to spectroscopic and other investigations carried out during the pre-Mariner period.

6 Venus in the Twentieth Century: Spectroscope and Theory

So far as stellar astronomy is concerned, the spectroscope is all-important. Without it, our knowledge of the stars would indeed be meagre. The Moon, and planets, however, shine by reflected sunlight, and with the Moon all we receive is a very enfeebled solar spectrum. However, Venus is rather more promising, since its atmosphere can leave its imprints upon the reflected spectrum; and at an early stage it was hoped that traces of oxygen and water vapour would show up. Preliminary work supported this suggestion. During the transits of 1874 and 1882, Tacchini and Riccò in Italy,[1] and Young in America,[2] found what they took to be certain evidence of water vapour; and where there is water vapour, there is an excellent chance of finding free oxygen. Scheiner, summing up the situation in 1894, wrote that 'there can therefore be no doubt that the atmosphere of Venus exerts an absorption similar to our own, and hence the nature of the two atmospheres must be similar'.[3]

Unfortunately, this early work proved to be far from reliable. The problem was clearly very difficult, since the spectrum of Venus is not easy to interpret. There is an additional complication, too; the rays of light have to pass through the Earth's atmosphere before reaching us, so that lines due to oxygen and water vapour in our own air will be present as well. Lines of this kind are termed 'telluric lines'. The main hope of disentangling the Cytherean lines from the telluric ones lay in the famous Doppler effect. When a body is approaching, its spectral lines are shifted towards the short-wave or violet end of the band; when the body is receding, there is a red shift. When Venus is moving away from the Earth, the telluric lines of oxygen will be unaffected, but the Cytherean ones will be shifted towards the red—so that the original lines should become double, or at least show a definite broadening. Similarly, when Venus is approaching, the doubling or broadening will be to the violet side of the original lines.

In theory, this method is perfectly sound, and in 1921 Slipher attacked the problem from the Lowell Observatory.[4] He failed completely. So did S. B. Nicholson and W. St. John, at Mount Wilson, in the following year.[5] In 1932 Adams and Dunham at Mount Wilson tried again, using improved equipment fitted to the 254-cm Hooker reflector, but once again the Cytherean oxygen refused to show itself.[6] Instead, Adams and Dunham detected infra-red absorptions which they could not at first identify, but which they subsequently found to be due to carbon dioxide.[7]

This was not only unexpected, but also rather disconcerting. For one thing, carbon dioxide produces a marked 'greenhouse effect', and blankets in the Sun's heat, so that a dense carbon-dioxide atmosphere must result in a high surface temperature. The new methods used by Adams and Dunham gave evidence of a state of affairs very different from that pictured thirty-five years earlier by Johnstone Stoney, who had concluded that the Cytherean atmosphere was moisture-laden and an almost complete copy of the Earth's.[8]

Slipher and Adel estimated that the amount of carbon dioxide in the atmosphere of Venus was equal to a layer 3·2 kilometres in thickness at a standard atmospheric pressure and temperature, compared with a mere 9 metres for the Earth.[9] Then, in November 1959, two American observers, Commander Ross (the pilot) and C. B. Moore, went up in a balloon to study Venus, and reported unmistakable signs of water vapour.[10] They stated that 'the measured Venus water is about four times more water than lies in our own stratosphere ... and is about the same as lies above high-level clouds on the Earth. We presume that there is much more water in the atmosphere of Venus below the cloud level.'

This was all very well, but there was still no definite information about the extent of the Cytherean atmosphere, or its structure. An early estimate by Watson gave 88 kilometres,[11] but was of historical interest only; Spencer Jones maintained that the atmosphere was 'likely to be somewhat less extensive than that of the Earth';[12] McEwen went to the other extreme, giving a depth of anything up to 1600 kilometres! Moreover, for obvious reasons only the top of the cloud-layer was accessible to spectroscopic analysis. In 1960 Brian Warner, then at the University of London Observatory, examined measurements of the spectrum of the night side of Venus made at the Crimean Astrophysical Observatory by N. A. Kozyrev, and announced confirmation of nitrogen bands, plus many apparent coincidences with lines due to oxygen, both neutral and singly ionized.[13] Yet there seemed no doubt that carbon dioxide was the main constituent, and there were suggestions from V. A. Firsoff in England[14] and N. Barabashov in the USSR[15] that magnetic effects might lead to a tendency for the carbon dioxide to rise above the free oxygen.

In 1937, following a visual and photographic study, R. Wildt put forward the suggestion that the clouds might be due to formaldehyde, a combination of carbon, hydrogen and oxygen (CH_2O), formed under the influence of ultra-violet light emitted by the Sun.[16] Though pure formaldehyde is colourless and unclouded, the slightest trace of water vapour added to it immediately produces a dense white cloud, and the molecules combine to form droplets of 'plastics'.[17] However, searches for formaldehyde bands in the ultra-spectrum of Venus proved to be fruitless.[18] Alternatively, there was a suggestion from H. Suess that the cloud layer was composed of salts such as sodium chloride and magnesium chloride, produced by the drying-up of former oceans on the planet.[19] We now know that this idea was very wide of the mark.

Neither was there much information about the Cytherean wind system. In 1908 A. W. Clayden had suggested that convection currents should result in thin and scattered cirrus clouds, floating in the upper atmosphere and covering Venus with a delicate filmy network.[20] Ross believed the clouds to be in violent motion;[21] so did Dollfus[22] and Kuiper.[23] It had long been believed that the Cytherean atmosphere contained particles of some sort. Lowell had been among the first to propose this,[24] and it was supported in 1937 by a long study carried out by the Russian astronomer B. Gerasimovič,[25] but the nature of the particles remained a puzzle. There was some supporting evidence from the observations of prolongations of the cusps, which had been beautifully photographed by Tombaugh in 1950.[26] Measures were made by W. Rabe[27] and by F. Link;[28] Link believed the prolongations to be due to a layer of fine dust held in suspension in the upper atmosphere of the planet. V. V. Sharonov found the horizontal refraction on Venus to be 20″,[29] and interpreted this low value as being due to the presence of a thin translucent cloud layer.

Meantime, continued efforts were being made to determine the rotation period by spectroscopic methods. The first work in this field was carried out in 1900 by the Russian astronomer A. Belopolsky,[30] using spectroscopic equipment together with the 76-cm refractor at the Pulkovo Observatory, Leningrad. If Venus is rotating on its axis, then (assuming a more or less normal axial inclination) one limb will be approaching us while the other is receding; therefore the approaching limb will show a violet Doppler shift, and the receding limb a red one. Given a reasonably quick rotation these shifts should be detectable, and Belopolsky arrived at a rotation period of 24h 42m, which he later amended to 35h.[31] Other spectroscopic determinations made in the following years were contradictory, and it began to look as though the rotation must be slow.

Some comments made in connection with this particular problem were peculiar. There was, for instance, the view expressed by M. W. Ovenden, then Secretary of the Royal Astronomical Society. Referring to the spectroscopic method, he stated: 'The reason it fails on Venus is that we do not see the surface of Venus at all. The whole planet is covered with dense white cloud.'[32] He also maintained that no large lunar-type craters could exist on Earth, as they 'would, in a few million years, be rubbed away by friction with the atmosphere as the Earth rotates underneath it'.[33] This seems to be a reversion to the old 'proof' that the Earth cannot be spinning round, as the result would be a permanent gale!

Then, in 1958, R. S. Richardson at Mount Wilson attacked the problem once again, and produced results which were of tremendous value, even though they were not conclusive.[34] Richardson pointed out that if the rotation period were as short as 22 hours, the globe of Venus should be appreciably flattened at the poles—but this was not so, indicating that the rotation period must be longer than that of the Earth. He found the rotation to be so slow that the Doppler effects were masked by inevitable errors of measurement. He interpreted his result in three different ways: (1) The direction of revolution is retrograde (i.e. east to west) with a period between 8 and 46 days, a statement with one chance in two of being correct; (2) The period is longer than 14 days direct, or longer than 5 days retrograde, a statement with sixteen chances in seventeen of being correct; (3) The period is longer than 7 days direct or longer than $3\frac{1}{2}$ days retrograde, a statement with 134 chances in 135 of being correct.

Temperature measurements were also made. Because of the lesser distance of Venus from the Sun, it was originally thought that the temperature must be high even at the top of the cloud layer, but the first really detailed study, made between 1923 and 1927 by E.

Pettit and S. B. Nicholson,[35] did not give any such result.

Pettit and Nicholson used thermocouples on the 254-cm Hooker reflector at Mount Wilson. Fundamentally, a thermocouple consists of a circuit made up of two differently constituted wires soldered end to end. If one of the joins is warmed while the other is kept at a constant temperature, an electric current will be set up in the circuit, and the amount of the current is a key to the rise in temperature responsible for it. In their more detailed paper, Pettit and Nicholson gave a value of −38°C for the bright or day side of Venus and −33°C for the dark or night side.[36] At the Crimean Astrophysical Observatory, N. A. Kozyrev made some measurements and arrived at a value of −90°C,[37] while in 1956 W. M. Sinton and J. Strong gave −40°C for both the day and night hemispheres.[38] Of course, this was no reliable key to the temperature of the actual surface—and until new techniques were introduced, in the 1960s, speculation was rife.

Moreover, the idea of life on Venus was taken quite seriously at least in some quarters during the period when Lowell's Martian canals were in vogue. C. E. Housden, a strong supporter of Lowell's theories, wrote a book about Venus in 1915,[39] and put forward some ideas which were, to put it mildly, rather extreme. He held that the 225-day rotation period is valid, so that convection currents are set up between the day and night hemispheres; deposits of ice and snow are formed just inside the dark hemisphere, and glaciers drift back into the sunlight, enabling the local inhabitants to pump the water back along conduits—these conduits being, of course, the linear streaks shown on Lowell's map! It is significant that though the august periodical *Nature* carried a review of the book,[40] Housden's theories were merely criticized rather than ridiculed.

Next came Svante Arrhenius, a Swede whose scientific work was good enough to earn him a Nobel Prize. In a book published in 1918,[41] he gave a vivid and highly attractive picture of Venus, which he pictured as a world rather in the state of the Earth more than 200 million years ago—in the Carboniferous Period, when the Coal Forests were being laid down and the most advanced life-forms were amphibians; even the great dinosaurs lay well in the future. Arrhenius' description of Venus is of historical interest, and is worth quoting at some length:

'The average temperature there is calculated to be about 47°C. . . . The humidity is probably about six times the average of that of the Earth, or three times that in the Congo, where the average temperature is 26°C. The atmosphere of Venus holds about as much water vapour 5 kilometres *above* the surface as does the atmosphere of the Earth *at* the surface. We must therefore conclude that everything on Venus is dripping wet. The rainstorms, on the other hand, do not necessarily bring greater precipitation than with us. The cloud-formation is enormous, and dense rain-clouds travel as high up as 10 kilometres. The heat from the Sun does not attack the ground, but the dense clouds, causing a powerful external circulation of air which carries the vapour to higher strata, where it condenses into new clouds. Thus, an effective barrier is formed against horizontal air-currents in the great expanses below. At the surface of Venus, therefore, there exists a complete absence of wind both vertically, as the Sun's radiation is absorbed by the ever-present clouds above, and horizontally, due to friction. Disintegration takes place with enormous rapidity, probably about eight times as fast as on Earth, and violent rains carry the products speedily downhill, where they fill the valleys and the oceans in front of all river mouths.

'A very great part of the surface of Venus is no doubt covered with swamps, corresponding to those on the Earth in which the coal deposits were formed, except that they are about 30°C warmer. No dust is lifted high into the air to lend it a distinct colour; only the dazzling white reflex from the clouds reaches the outside space and gives the planet its remarkable, brilliantly white lustre. The powerful air-currents in the highest strata of the atmosphere equalize the temperature difference between poles and equator almost completely, so that a uniform climate exists all over the planet analogous to conditions on the Earth during its hottest periods.

'The temperature on Venus is not so high as to prevent a luxuriant vegetation. The constantly uniform climatic conditions which exist everywhere result in an entire absence of adaptation to changing exterior conditions. Only low forms of life are therefore represented, mostly no doubt belonging to the vegetable kingdom; and the organisms are of nearly the same kind all over the planet. The vegetative processes are greatly accelerated by the high temperature. Therefore, the lifetime of organisms is probably short. Their dead bodies, decaying rapidly, if lying in the open air, will fill it with stifling gases; if embedded in the slime carried down by the rivers, they speedily turn into small lumps of coal, which later, under the pressure of new layers combined with high temperature, become particles of graphite.

'. . . . The temperature at the poles of Venus is probably somewhat lower, perhaps by about 10°C, than the average temperature on the planet. The organisms there should have developed into higher forms than elsewhere, and progress and culture, if we may so express it, will gradually spread from the poles toward the equator. Later, the temperature will sink, the dense clouds and the gloom disperse, and some time, perhaps not before life on the Earth has reverted to its simpler forms or has even become extinct, a flora and a fauna will appear, similar in kind to those which now delight our human eye, and Venus will then indeed be the "Heavenly Queen" of Babylonian fame, not because of her radiant lustre alone, but as the dwelling-place of the highest beings in our Solar System.'

Arrhenius was not alone in his views. Similar ideas were expressed by F. W. Henkel,[42] G. Zech[43] and C. G. Abbot,[44] among others. On the other hand there were also those who regarded Venus as an arid dust-bowl, and F. E. Ross suggested that the surface might be of a uniform red-yellow hue.[45] Estimates of the surface temperature varied; 'over 50°C' by Adel in 1937[46] and 1952, 'high' by R. Wildt in 1940[47] and G. Herzberg in 1951,[48] 30°C by N. Kozyrev in 1954[49] and so on. A rather unexpected picture was proposed by Sir Fred Hoyle in 1955.[50] Instead of Arrhenius' wet world or the increasingly popular dust-bowl, he suggested that there might be oceans of oil! His reasoning was as follows:

'Suppose an enormous quantity of oil were to gush to the Earth's surface; what would the effect be? The oil, consisting as it does of hydrocarbons, would proceed to absorb oxygen from the air. If the amount of oil were great enough all the oxygen would be removed. When this happened the water vapour in our atmosphere would no longer be protected from the disruptive effect of ultra-violet light from the Sun. So water vapour would begin to be dissociated into separate atoms of oxygen and hydrogen. The oxygen would combine with more oil, while the hydrogen atoms would proceed to escape altogether from the Earth out into space. More and more of the water would be dissociated and more and more of the oil would become oxidized.

The process would only come to an end when either the water or the oil became exhausted. On the Earth it is clear that water has been dominant over oil. On Venus the situation seems to have been the other way round, the water has become exhausted and presumably the excess of oil remains—just as the excess of water remains on the Earth.

'This possibility has an interesting consequence. In writing previously about these clouds I said that the only suggestion that seemed to fit the observations was that the clouds are made up of fine dust particles. To this suggestion we must now add the possibility that the clouds might consist of drops of oil—that Venus may be draped in a kind of perpetual smog. . . .'

And with regard to the slow axial rotation:

'It is thus reasonable to suppose that the slowing-down of Venus can be explained by the friction of tides—if Venus possesses oceans, but not I think otherwise. Previously the difficulty was to understand what liquid the oceans were made of. Now we see that the oceans may well be oceans of oil. Venus is probably endowed beyond the dreams of the richest Texas oil-king.'

In view of the present world situation, this would make Venus an attractive target! But at about the same time (1955), two eminent American astronomers, F. L. Whipple and D. H. Menzel, came out with a picture which was different again. This time the Cytherean oceans were nothing more nor less than ordinary water, with clouds made up of H_2O.[51] The first announcement of the theory was given in a paper read to the American Astronomical Society's convention at Ann Arbor in June 1954.

The H_2O theory was based essentially on a series of measurements of the polarization of the light of Venus, made in 1929 by the great French astronomer Bernard Lyot.[52] Among the substances with which he was able to experiment, Lyot found that only water droplets agreed reasonably well with the variation of polarization with scattering angle on Venus. In their paper, Whipple and Menzel wrote:[51]

'Lyot's polarization measures indicate that water droplets fit his data satisfactorily. The droplets of the Venusian clouds are uniform in dimension and large for high-level airborne dust. No one has been able to suggest, in place of water droplets, a likely substitute material that would both be available and agree with

the polarization and other reflection characteristics observed on
the clouds.'

They therefore suggested that the surface of Venus was likely to be
completely covered with water.

On this theory, a thick atmosphere consisting largely of carbon
dioxide could not exist upon an earth-like planet with continents
protruding from oceans of water; the carbon dioxide would be fixed
in the rocks to form carbonates, because of its chemical reaction
with silicates in the presence of water. If protruding land masses
were virtually absent, however, the fixation of carbon dioxide could
not continue after the formation of a thin buffer layer of carbonates,
and this was the reason for assuming Venus to be essentially
oceanic. There was one rather bizarre corollary. Presumably the
atmospheric CO_2 would have fouled the oceans, giving Venus seas
of nothing more nor less than soda-water (though, as was pointed
out by various writers, there was not likely to be any whisky to mix
with it!).

Actually, the idea of Cytherean oceans was not new. As long ago
as 1873 A. Safarik had tried to explain the Ashen Light by
phosphorescent water (see Chapter 7); in 1924 W. H. Pickering had
suggested that there might be seas,[53] and H. Jeffreys had written
that 'Venus may have an ocean with shallow seas, much like ours'.[54]
On the other hand, H. C. Urey maintained that in all probability
Venus once had extensive oceans, perhaps with life, but by now all
the water had vanished, leaving Venus a sterile world.[55] He stated
that 'the presence of carbon dioxide in the planet's atmosphere is
very difficult to understand unless water were originally present,
and it would be impossible to understand if water were present
now.'

It is, of course, true that in very early times our own Earth had an
atmosphere which we would find unbreathable; there was a great
deal of carbon dioxide, and little free oxygen. Free oxygen became
plentiful only when plants spread on to the lands, and the process of
photosynthesis became all-important. If Venus were a watery
world, why should not life have begun there, just as it did in the seas
of Earth? The prospect was alluring, and at the time when Whipple
and Menzel put forward their theory it could certainly not be ruled
out. It followed that over the ages Venus could well have produced
more advanced life-forms. Also in 1955, some interesting
speculations were made in the USSR by G. A. Tikhov,[56] who

studied what he termed 'astrobotany' and 'astrobiology'. He wrote:

'Now already we can say a few things about the vegetation of Venus. Owing to the high temperature on this planet, the plants must reflect all the heat rays, of which those visible to the eye are the rays from red to green inclusive. This gives the plants a yellow hue. In addition, the plants must radiate red rays. With the yellow, this gives them an orange colour. . . . Our astrobotanical conclusions concerning the colour of vegetation on Venus find certain confirmation in the observations of Academician N. P. Barabashov. He found that in those parts of Venus where the Sun's rays possibly penetrate the clouds to be reflected by the planet's surface, there is a surplus of yellow and red rays. Barabashov believes that the surface of Venus is to a certain degree specular, and that the yellow and red rays pass more easily through the clouds than do the rays at the blue end of the spectrum. I in my turn wish to add that here a certain part may be played by the vegetation of Venus. Thus we get the following gamut of colours: on Mars where the climate is rigorous the plants are of blue shades. On Earth where the climate is intermediate the plants are green, and on Venus where the climate is hot the plants have orange colours.'

Tikhoff supported the general idea that Venus today is rather like the Earth must have been a hundred million or more years ago.

It has seemed worthwhile citing these various opinions at some length, if only to stress how diverse the opinions of different astronomers were. Moreover, our ignorance of the surface relief was complete. Some comments upon the possibility of Cytherean mountains were made by H. C. Urey in 1956.[57] According to Urey:

'It must be expected that mountain building is associated with energy sources within the planet such as those due to radioactivity and the formation of planetary cores. All such sources become smaller with time and eventually mountain building must cease. Also the smaller the planet the less vigorous is the mountain building activity and the sooner will this activity cease. Thus the Moon never produced any volcanoes or terrestrial type mountains. Probably Mars did not do so either. Venus may have produced its mountains in the past and they may have been eroded to sea level perhaps in the distant past and it may be that only the Earth has sufficient size to continue this activity at the present time.'

In that period there was certainly no thought that Mars might have giant volcanoes far higher than any on the Earth. Another hypothesis which has turned out to be very wide of the mark was the so-called aeolospheric model, due to E. Öpik in 1961;[58] Öpik believed that the deep atmosphere of Venus is filled with dust raised by winds near the solid surface.

An entirely new approach to the problem of the rotation of Venus was made in 1956 by J. F. Kraus. In February of that year he began to search for radio emission from Venus, using equipment operating at a wavelength of 11 metres, set up near Columbus, Ohio. By May he was convinced that such emission had been found,[59] and described the signals as having burst-like characteristics not unlike thunderstorm atmospherics; indeed, he regarded it as quite possible that they were due to thunderstorms above Venus. The bursts were of short duration, lasting no more than the fraction of a second. In July, Kraus reported[60] a second type of signal on the same wavelength, of greater duration, amounting to a second or more, and bearing a slight resemblance to signals from an Earth radio station. He also claimed a definite relationship between outbursts on the Sun, the Venus signals, and radar echoes from the Moon.

Kraus' third paper led on to an estimate of the rotation period by radio methods.[61] The strength of the signals was found to vary in a regular manner, and Kraus accordingly wrote as follows:

'Although the mechanism producing the fluctuations in the signals from Venus is not known, a simple, plausible explanation might be the following. It is likely that Venus has an ionosphere at least as dense as that of the Earth. Radio waves originating on or near the surface of the planet and reaching the Earth would most readily penetrate this ionosphere in the general vicinity of the planet's central point. In this vicinity the ionosphere would appear more transparent to signals travelling earthward, so that, in effect, the ionosphere would have a hole. Assuming that the hole remains relatively fixed in position with respect to the central portion of the planet's disk, the sources would be observed only as they pass, by virtue of the planet's rotation, beneath the hole, each producing a peak of activity as it passes by.'

Kraus gave a rotation period of 22h 17m, with a possible uncertainty of only ten minutes either way. It is rather ironic that his result eventually proved to be in error by more than eight *months*!

And later examination showed that there had been serious misinterpretations in Kraus' work, which is now of historical interest only.

To recapitulate: at the end of the nineteenth century Venus was known to have an extensive atmosphere, but its composition was open to speculation. The early spectroscopic observations showed Venus to have a particularly bright spectrum, but the early observations searching for oxygen and water vapour failed to detect any Cytherean absorption lines. Later observations at Mount Wilson could only set upper limits of ≤3 atmospheres and <1 p.mm for oxygen and water vapour respectively.[62]

The development of new photographic emulsions allowed spectroscopy to be extended to the near infra-red wavelengths. The observations made by Adams and Dunham confirmed the presence of CO_2 absorption features in the atmosphere.[63] They suggested that the abundances of oxygen and water vapour in this region were by at least a factor of 50 less than in the terrestrial atmosphere.

Between 1924 and 1928 observations of many stars and planets, including Venus, were carried out at Mount Wilson, using the Hooker reflector to focus the radiation on to vacuum thermopile detectors.[64,65] The scans of Venus revealed little variation in the thermal emission between the dark and illuminated sides of the planet, and respective black-body temperatures of 240°K and 235°K were defined.

By the Second World War it seemed that Venus had an atmosphere containing large quantities of carbon dioxide but little oxygen or water vapour. Its surface was obscured by clouds which scattered light in a similar way to water-droplets, and emitted thermal radiation at an effective temperature of 240°K. But of what were these clouds composed? What was the structure of the atmosphere beneath the clouds? Equally, at this stage we had little knowledge of the radius of the solid planet, its rotation period or its axial inclination to the plane of the ecliptic. To answer these fundamental problems it required advances in electronics and detector technology.

Rotation and Radius

As we have seen, the rotation period of Venus is very difficult to estimate. In most situations, it is usual to determine a value from studies of optical features on the planetary disk. But at visible wavelengths there is nothing to be seen in the Cytherean clouds; in

1667 Cassini favoured a period near 24 hours, but the transient and generally illusory nature of the features recorded led to contradictory results. In 1890 Schiaparelli postulated a Sun-synchronous rotation. Many of the early measurements (see table, page 167) gave a wide range of estimates.

It was not until 1961 that the first reliable radar contact with Venus was made by the team at the Lincoln Laboratory in the United States (the success first claimed in 1958 proved to be premature). The initial aim was to make a very accurate measurement of the distance of Venus, which would lead to a determination of the length of the astronomical unit or Earth-Sun distance; radar mapping of the surface would follow later. Meantime, observations at wavelengths of 12·5 cm were obtained, and all the measurements suggested a very slow rotation period. Goldstein and Carpenter[66] and Carpenter[67] found that the period was 250±40 days in a retrograde sense, with a rotational axis inclined at nearly 180 degrees to the orbital plane of the planet.

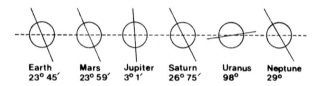

Earth	Mars	Jupiter	Saturn	Uranus	Neptune
23° 45′	23° 59′	3° 1′	26° 75′	98°	29°

Fig. 10   Axial inclinations of the planets.

Subsequent observations (see table, page 167) refined this period to 243±10 days, which is very close to the period of 243·16 days corresponding to synodic resonance with the Earth's orbit, so that Venus would present the same face to the Earth at each inferior conjunction after four complete revolutions relative to the Earth.

However, at the cloud tops there was a different story. Photographs of the planet at ultra-violet wavelengths revealed large-scale dark markings that appeared to move round the planet with a period of approximately 4 days.[68] Subsequently studies by many observers (see table, page 167) confirmed a period of this order. A study of features for a near-lifetime of two days suggested that the rotational period ranged from 3·8 to 4·2 days.[69]

Spectroscopic measurements of the Doppler shift of the Fraunhofer lines and near-infra-red absorption lines in the radiation reflected from the planet were consistent with retrograde zonal winds of 83 ms^{-1} at equatorial latitudes.[70] These values have now been found to be consistent with the observations made from the more recent space-craft missions to Venus.

Although the radius and sphericity of Venus at the level of the cloud-tops were known quite accurately, it required observations at radio and microwave lengths to probe the ubiquitous cloud, and determine the shape and structure of the surface of the planet. Radii decreasing from 6150 km at ultra-violet wavelengths to 6120 km at visible wavelengths were derived from photographic plates, and the analysis of the occultation of Regulus by Venus in 1959. However, the early microwave measurements gave a radius of 6057±55 km for the surface. This value is about 60 km below the clouds, and it was later confirmed by radar techniques, strengthening the evidence for an extensive, hot lower atmosphere.

Rapid advances in radar technology have now led to detailed topographical mapping of the surface of Venus, both by Earth-based radio telescopes and by satellite-mounted radar systems. Observations made at 3·8 and 70 cm displaced brightness variations with a horizontal resolution of 200 km. Relative differences in brightness between these correspond to large-scale variations of 6 km in surface altitude.[71] Similar equatorial topography and a mean equatorial radius of 6050±0·5 km were later derived by Campbell *et al.*[72] Maps of limited areas and giving a horizontal resolution of 10 to 20 km revealed that Venus was cratered, and Rumsey *et al.* identified a 160-km diameter crater with a depth of about 500 metres.[73] These studies of the surface of Venus have been extended by satellite observations, which provide global coverage of the planet, which is now represented by a mean radius of 6052 km.

The Clouds of Venus

The composition of the Cytherean clouds has been the centre of many discussions over the past few decades, but until about 1971 our knowledge was still in a very confused state. At that stage, no space-probe had actually analysed the cloud particles. The list of possible candidates was quite long. They included water-droplets; ice crystals; hailstones; derivatives of methane, ethane and benzene; dust; snowflakes; ammonium chloride and ice-HCl solution; and volcanic products.

Many of the initial speculations were simply based upon the

conception that the Earth has water and ice clouds, so they could presumably exist also on Venus—which was still thought of as the twin of our own planet. However, as we have seen, the Earth and Venus appeared to have major differences, which became more marked as we learned more about our planetary neighbour. Many of the initial analyses were based upon attempts to fit the spectrum of the sunlight reflected by the cloud layers.[72] Since this radiation is reflected at different levels in the clouds, there is always the problem that clouds of different compositions and varying sizes of particles are confusing the observations.

A further complication is that the water measurements made by the earlier Venera space-craft were really consistent with clouds of water ice or strong solutions of HCl. But the ground-based spectroscopic observations were difficult to reconcile with this conclusion. However, a potentially valuable way of resolving the problem is from analyses of the polarized radiation reflected by the clouds. This characteristic of the reflected radiation is very sensitive to the size, shape and composition of the cloud particles, which can be determined from observations of the entire 180° of phase angle, because Venus is an inferior planet. Furthermore, all the radiation is reflected by the cloud-tops, so that change in the cloud composition at other levels in the atmosphere cannot confuse the issue.

From this type of study, it was found that the cloud particles were spherical, with a radius of $1 \cdot 05 \pm 0 \cdot 1$ µm, with a refractive index of $1 \cdot 44 \pm 0 \cdot 015$.[73,74] Clearly these particles were not water-droplets, since the refractive index of water would be $1 \cdot 33$. The tops of the clouds were at a pressure level of 50 ± 25 mb, where the temperature is about 240°K. It was found that a 75 per cent solution by weight of sulphuric acid droplets was the most likely candidate to satisfy all the observations.[75,76] Certainly the strong hydroscopic nature of H_2SO_4 would explain the low water vapour abundance measured in the neighbourhood of the cloud-tops. Sulphuric acid droplets in the upper clouds also explained many other of the previously perplexing observations of the Cytherean clouds. However, there was still the problem of providing a suitable explanation of the ultra-violet constraints. Absorption in the dark regions was thought to be HBr (hydrogen bromide), or even elemental sulphur. Sulphur was also postulated as part of a chemical cycle involving COS and H_2SO_4, and at this stage it was thought that its rôle in the troposphere of Venus could be comparable with that of water in its vapour and liquid states in the terrestrial atmosphere. As a consequence, there was the possibility of highly corrosive rainstorms in the Cytherean atmosphere!

But our knowledge of conditions beneath the clouds was still limited. Similarly, the precise composition of the atmosphere remained unknown. The analyses of the Earth-based telescopic observations were strongly affected by the effects of clouds, which also prevent radiation penetrating through from below. By the early 1970s we were sure that carbon dioxide was the major constituent of the atmosphere, with a concentration of from 93 to 97 per cent. Nitrogen was thought to make up 2 to 5 per cent, while the oxygen content was considered to be less than 0·7 per cent. Also, minor constituents present were CO, HCl, HF and H_2O.[77] Yet the concentrations of these constituents would be affected by the local chemistry, the cloud formation processes and the atmospheric motions. We needed to penetrate the clouds and measure the atmospheric properties at all levels before these questions could be answered in a quantitative manner.

The Microwave Brightness Problem

The uniform temperatures of 240°K inferred from early measurements of the thermal emission of Venus, together with the high reflectivity of 79 per cent, immediately suggested that the clouds behaved like a reflecting layer, with their tops high in the Cytherean troposphere. Consequently, the high brightness temperatures derived from the first measurements of the microwave emission from Venus obtained during the 1956 inferior conjunction aroused considerable interest.[78] If thermal in origin, the intensity of radiation absorbed at 3·15 cm in the two separate experiments indicated temperatures of 620±110 and 560±73°K for the limiting region. Further measurements at 9·4 cm suggested a temperature of about 600°K.

During the next few years, several studies carried out in the United States and the Soviet Union extended the microwave coverage of Venus from 0·4 to 40 cm. Between 5 and 15 cm the brightness temperature was found to remain virtually constant at 650 to 700°K. To what do these temperatures refer? Are we really observing a hot surface on Venus?

Of course, we now know the answer; but nearly twenty years ago a whole range of possible explanations came forward. As an alternative to a hot surface, it was suggested that the radiation was generated in the Cytherea ionosphere, and that the atmosphere was cooler and less extensive. The observations from the Mariner 2 space-craft microwave experiment supported a thermal rather than an ionospheric source. Furthermore, we now know that Venus' magnetic dipole is extremely weak, and only 1/30 of the terrestrial value, which reduces the likelihood of an extensive ionosphere on the planet.

After these lengthy debates we now had a more complete picture of a dense, massive atmosphere, over a hot surface. Already, at this stage, Venus was showing important differences from the Earth. With the Space Age, we could come to an exciting, detailed explanation of Venus, and a more detailed understanding of the situation prevailing beneath the clouds.

7 The Ashen Light

When the Moon is a crescent, high enough to be seen against a fairly dark background, the unlit 'night' part of the disk may often be seen shining with a faint light. This phenomenon is known to country folk as 'the Old Moon in the Young Moon's arms', and has been known for many centuries. Leonardo da Vinci was probably the first to give a correct explanation for it. It is due simply to light reflected from the Earth on to the Moon.

A similar effect has been seen on Venus, though naturally it cannot have the same origin. Apparently it was first noted on 9 January 1643, by Giovanni Riccioli,[1] a Jesuit professor at Bologna. Riccioli's chief claim to fame is that he drew up a map of the Moon in 1651 in which he named the principal craters in honour of eminent men and women instead of keeping to the feeble geographical analogies of his predecessors. Even though he cannot be compared with men such as Huygens and Cassini, he was a competent astronomer (even if he did decline to accept the heretical idea that the Earth is in orbit round the Sun). Riccioli himself regarded this particular observation as being rather dubious.

The Ashen Light, as the faint luminosity is called, was next recorded in 1714 by Derham, and subsequently by many observers. An excellent summary of the reports from 1643 to 1900 has been given by R. M. Baum.[2] Earlier accounts have been given in *Nature*[3] and in a paper by A. Safarik, which deals with the observations made before 1873.[4]

William Derham, Canon of Windsor, saw the Ashen Light about 1714. In his words: 'This sphæricity, or rotundity, is manifest in the Moon, yea in Venus too, in whose greatest falcations the dark part of their globes may be perceived, exhibiting themselves under the appearance of a dull and rusty colour.'[5] From a footnote to the third edition of Derham's book, published in 1719, it seems that he saw the Light frequently.

There is little point in giving the details of all the later observations, and in fact it would be tedious to do so, but a few of the more notable earlier records are worth listing: Kirch in 1721,[6] Mayer in 1759,[7] Hahn in 1793,[8] William Herschel around 1790 to 1795,[9] Schröter in 1806,[10] Harding in 1806,[11] Pastorff many times around 1822,[12] Guthrie in 1842,[13] Berry in 1862,[14] Prince in 1863,[15] Engelmann in 1865,[16] Petty in 1868,[17] Browning in 1870,[18] Safarik in 1870,[4] Winnecke in 1871,[19] Van Ertborn in 1876,[20] Zenger in 1876,[21] Webb in 1878,[22] Zenger again in 1883,[23] Lohse and Wigglesworth in 1886[24] and McEwen in 1895.[25] Schröter saw the Light only once, but on that occasion (14 February 1806) he described it as being very clear, with the limb of the night hemisphere brighter than the central part. On 25 September and 6 November 1871 A. Winnecke, of Karlsruhe, reported the whole disk, 'the dark side bathed in a pale greyish light; quite distinct, and free from suspicion of illusion'. Winnecke was undoubtedly a good observer, and is well remembered because of his discovery of a periodical comet which bears his name.

After 1890 or thereabouts it seems that the Ashen Light has been glimpsed by almost every serious observer of Venus—apart from Barnard, who was never able to detect it.[26] Normally the Light is seen only when Venus is a thin crescent. In June 1895 Leo Brenner, at the Manora Observatory, claimed to have seen it during the gibbous phase[27]—but as we have noted, Brenner's work cannot be regarded as reliable. However, the Light was again reported at a greater phase than usual in 1951 by T. L. Cragg and J. C. Bartlett.[28] On one occasion in 1927 C. S. Saxton saw the whole disk 'not as a ring, but as a very faint whiteness or mistiness against the blue sky'.[29] It is clear that the observations are of various types.[30]

The term 'Ashen Light' should properly be restricted to the faint luminosity of the night part of the disk, and should not be extended to cover reports of the night side being seen as *darker* than the background. This latter can be due only to contrast effects. (Flammarion once suggested that it might be caused by the subdued background brightness of the sky owing to the presence of the Zodiacal Light, but this idea seems to be untenable.)

As an example of a relatively modern description of the Ashen Light, it is worth quoting M. B. B. Heath, a very experienced observer who spent many years in studying Venus.[31]

'When seen in daylight, the unilluminated portion of the disk was invariably noted as being darker than the surrounding sky, the darkness being frequently more prominent near the terminator and gradually shading off into invisibility at some point near the line joining the cusps, but sometimes extending over the whole or nearly the whole disk. After sunset, at some rather indefinite time, the dark side has been noted as brighter than the outside sky, sometimes showing a dull red or brownish tint; usually this dim glow is seen over the whole unlit area of the planet. On two occasions it has been noted to be generally mottled, or of uneven brightness.'

Heath's private observatory, at Kingsteignton in Devon, was equipped with an excellent 23-cm reflector.

One of the present authors (Moore) has been observing Venus regularly ever since 1934, using mainly his 32-cm and 39-cm reflectors. The projection of the dark disk against the sky has never been seen with certainty, but the true Ashen Light, or faint luminosity of the night hemisphere, has been seen so often and so clearly that it is hard to dismiss as a mere illusion. To guard against contrast effects, it is virtually essential to block out the bright crescent by means of an occulting bar. One particularly striking view was obtained on 27 May 1980, just after sunset. The Ashen Light was so conspicuous that it looked almost like the earthshine on the Moon, with its brightest part near the edge of the disk.

8 The Ashen Light, 27 May 1980, 20h.20m. 39cm reflector, × 72 to 300. Patrick Moore. Seeing excellent (1). The crescent was hidden by an occulting bar to check that there was no contrast effect involved. The visibility of the dark side has been enhanced for the sake of clarity.

Genuine or not, the Ashen Light has been seen so frequently, and by so many experienced observers, that it has to be explained in some fashion. One suggestion, put forward by Franz von Paula Gruithuisen in the early nineteenth century, is perhaps worth recalling. Gruithuisen was an enthusiastic observer, who was concerned mainly with the Moon; unfortunately his imagination was so vivid that he tended to bring a good deal of ridicule on himself—for instance, he was convinced that he had discovered artificial structures on the Moon. With regard to Venus, he pointed out that the Ashen Light had been observed in 1759 and again in 1806, an interval of 47 terrestrial or 76 Cytherean years, and went on to say:[32]

> 'We assume that some (Cytherean) Alexander or Napoleon then attained universal power. If we estimate that the ordinary life of an inhabitant of Venus lasts 130 Cytherean years, which amounts to 80 Earth years, the reign of an Emperor of Venus might well last for 76 Cytherean years. The observed appearance is evidently the result of a general festival illumination in honour of the ascension of a new emperor to the throne of the planet.'

Later on, Gruithuisen modified this theory somewhat. Instead of a Cytherean Coronation, he suggested that the Light might be due solely to the burning of large stretches of jungle to produce new farm land, and added that 'large migrations of people would be prevented, so that possible wars would be avoided by abolishing the reason for them. Thus the race would be kept united.'

Not surprisingly, Gruithuisen found few supporters, so let us turn to other explanations, which may be more plausible even if less dramatic.

First, there is possible illumination by some other body. Venus has no satellite, and therefore the only possible candidate is the Earth, which would admittedly be brilliant as seen from Venus—or, more accurately, from above the cloud tops. In 1859 Rheinauer of Munich calculated that by Earth illumination the Ashen Light would be of about the fourteenth magnitude,[33] and in 1883 C. V. Zenger of Prague strongly supported the idea.[23] But, to be candid, any elementary calculation shows that it is absolutely inadequate, and is completely out of court.

Secondly, there is the phosphorescence theory. This was supported by William Herschel, who saw the Light several times around 1790, and by Schröter, who, as we have noted, saw it once only. A later supporter was H. Klein.[34] Again, however, the explanation is not only inadequate, but palpably so. We must also reject J. E. Pastorff's theory of a self-luminous atmosphere.[12] Neither has H. Vogel's idea[35] of 'a very extensive twilight'[35] anything to recommend it.[35] And even worse was a much later suggestion by Barker that the Light could be the reflection of a layer of ice on the surface of Venus.[36] Conditions there do not seem to be very well suited to the formation of ice sheets!

Safarik, author of a useful review paper in 1873, strongly supported the idea of shining oceans. He wrote:[4]

'The intense brightness of Venus, and particularly the dazzling splendour of her bright limb, is deemed by the late G. P. Bond and by Professor Zöllner, a competent authority in photometric matters, not to be explicable without assuming specular reflection on the surface of the planet. This Professor Zöllner supposed to be done by a general covering of water, and indeed if the faint grey spots of Venus, delineated in 1726 by Bianchini and rediscovered by Vico in 1838, are land, then nine-tenths at least of the surface of Venus are covered by sea. Should Venus be in a geologically less advanced state, viz. less cooled than our globe, a supposition rendered not improbable by her considerable size and her nearness to the Sun, then the present condition of Venus would be analogous to that of the Earth on the Jurassic Period, when large isolated islands were bathed by immense seas, blood-warm, and teeming with an abundance of animal life difficult to be conceived.

'The intensity of the phosphorescence of the sea, shown not infrequently by our tropical seas, gives us some idea of the intensity which this magnificent phenomenon could acquire under such unusual circumstances; and it is, I think, not unreasonable to expect that such a phosphorescence could be seen even at planetary distances. It would explain the fact that the edge of the dark hemisphere of Venus is seen brighter than its central part; for it is demonstrable by calculation and confirmed by observation (as in the case of the sea near the horizon, or the edge of the full moon), that a rough surface emitting diffused light is seen the brighter the more obliquely it is regarded.'

Coming on to modern times, there are various authorities who regard the Light as being due to nothing more than a contrast effect. To this school of thought belong the French observers D. Barbier[37] and A. Danjon.[38] Of course, it is always unwise to be dogmatic when discussing details near the limit of visibility—remember the canals of Mars!—but on the whole it does seem that the observations of the Light are too numerous, and too concordant, to be dismissed.

Much more promising are the theories which attribute the Light to electrical effects, or to Cytherean auroræ. The first suggestions of this kind seem to have been due to P. de Heen in 1872[39] and J. Lamp in 1887.[40] At first glance there is nothing improbable about them. Terrestrial auroræ are caused by charged particles from the Sun entering the upper atmosphere. Venus, considerably closer to the Sun than we are, might well be expected to have auroræ on a much grander scale. This would probably indicate the presence of a Cytherean magnetic field; and in 1955 some interesting results were announced by J. Houtgast of the Sonnenburgh Observatory at Utrecht in Holland.[41] Houtgast reasoned that if Venus is a powerful magnet, then when Venus lies more or less between the Earth and the Sun—that is to say, when near inferior conjunction—it should deflect the charged solar particles; and from studies of magnetic records over a period of forty-four years, and taking into account the variations in solar activity, he believed that such effects were measurable, so that the magnetic field of Venus was estimated as being five times stronger than that of the Earth. All this sounded highly convincing at the time; but we now know that Venus has to all intents and purposes no general magnetic field at all, in which case Houtgast's correlations must have been spurious.

At the Crimean Astrophysical Observatory, N. A. Kozyrev obtained spectrograms of the dark side of Venus,[42] using a quartz spectrograph together with the 125-cm reflector at the Observatory. He reported spectral bands due to ionized nitrogen, at 3914 and 4278 Ångströms, and commented that these lines are also seen in the spectra of auroræ. From these results, he calculated that the brightness of Venus' night sky should be about fifty times as great as that of the night sky of the Earth; and V. Fesenkov added that 'Thus, the luminescence of Venus' nocturnal sky is apparently analogous to our northern lights, and, therefore, has much greater energy of radiation than the ordinary night sky of our terrestrial atmosphere.'[43] Brian Warner made further analyses of Kozyrev's data,[44] and found evidence of spectral lines due to oxygen; he found no trace of the oxygen line at 5577 Ångströms which is seen in the spectrum of the Earth's airglow, and concluded therefore that the spectra really did apply to the light from Venus rather than that of our own atmosphere. In 1958 G. Newkirk, at the High Altitude Observatory in Colorado, obtained some spectrograms of the night side of Venus in which he was able to make a tentative identification with lines of ionized nitrogen, which went some way towards confirming Kozyrev's results.[45] Newkirk estimated that the night-glow of Venus was some eighty times brighter than that of the Earth. In the following year (1959) Newkirk and J. L. Weinberg were less successful, and took twenty spectrograms of the dark side of Venus without being able to identify any emission lines at all.[46] It may or may not be significant, however, that visual observations of the Ashen Light were much more positive in 1958, when Newkirk recorded the lines, than in 1959, when he did not.

If the Light really is due to electrical phenomena—perhaps to auroræ—then there could well be some correlation with activity on the Sun; it is near the time of solar maximum that terrestrial auroræ are brightest and most frequent. According to members of the Mercury and Venus Section of the British Astronomical Association,[47] the Light was particularly noticeable around 1957–8, when the Sun was at the peak of its cycle, and it was also conspicuous around the time of the solar maximum of 1980. On the other hand, we must be very wary of jumping to conclusions. During the solar maximum of 1968–9 there were few reports of the Ashen Light. Moreover, observational selection is bound to play a part; the Light is only observable when Venus is in the crescent stage, and observations of it are too few, too scattered, and in many cases too unreliable to provide conclusive data one way or the other. There may or may not be a connection with events on the Sun, but at the moment nothing of the sort can be proposed with any confidence at all.

Yet, in anticipating Part 2 of this book, we have now been able to show that there are violent electrical phenomena in the Cytherean atmosphere, with almost continuous lightning and a marked atmospheric glow. Whether or not this is responsible for the Ashen Light remains to be seen. All we can really say is that if the Light as reported by Earth-based observers is real, and is not due to mere contrast, then it is almost certainly electrical in nature.

8 The Phantom Satellite

Most of the planets in the Solar System are attended by satellites. Of the giant planets, Neptune has three satellites, Uranus five, and Jupiter and Saturn more than a dozen each. The smaller planets are less richly endowed. Mars has two dwarf attendants only a few kilometres in diameter, which may well be ex-asteroids which were captured by Mars in the remote past (though on this point opinions differ). The Earth, of course, has its Moon. Even Pluto seems to have a satellite, though, as we have seen, the nature of Pluto is problematical. This leaves us with only two solitary planets: Mercury and Venus.

Some of the satellites are large. The four brightest members of Jupiter's family (Io, Europa, Ganymede and Callisto) were detected in 1610, and were seen by Galileo with his primitive telescope. In 1655 Christiaan Huygens found Titan, the senior attendant of Saturn. The satellite problem was next taken up by G. D. Cassini after his move to Paris. Cassini was successful in finding four other members of Saturn's retinue (Iapetus, Rhea, Dione and Tethys). Then, in 1686, he made what was thought to be an equally startling discovery. The following is an extract from his observational journal:

'1686 August 18th, at 4.15 in the morning. Looking at Venus with a telescope of 34 feet focal length, I saw at a distance of 3/5 of her diameter, eastward, a luminous appearance, of a shape not well defined, that seemed to have the same phase with Venus, which was then gibbous on the western side. The diameter of this object was nearly one quarter that of Venus. I observed it attentively for 15 minutes, and having left off looking at it for four or five minutes; I saw it no more; but daylight was by then well advanced. I had seen a like phenomenon, which resembled the phase of Venus, on 1672 January 25, from 6.52 in the morning, to 7.02, when the brightness of the twilight caused it to disappear.

Venus was then horned, and this object, which was of diameter almost one quarter that of Venus, was of the same shape. It was distant from the southern horn of Venus a diameter of Venus on the western side. In these two observations, I was in doubt whether it was or was not a satellite of Venus, of such a consistence as not to be very well fitted to reflect the light of the Sun, and which in magnitude bore nearly the same proportion to Venus as the Moon does to the Earth, being at the same distance from the Sun and Earth as was Venus, the phases of which it resembled.'

It was then recalled that Fontana had seen something similar as far back as 15 November 1645,[1] but the question remained in abeyance for some time. Then, near sunrise on 23 October 1740, the satellite was recorded by James Short, the well-known instrument-maker. Short's account of it is interesting enough to be reproduced in full:[2]

'Directing a reflecting telescope, of 16·5 inches focus (with an apparatus to follow the diurnal motion) toward Venus, I perceived a small star pretty nigh upon her; upon which I took another telescope of the same focal distance, which magnified about 50 or 60 times, and which was fitted with a micrometer, in order to measure the distance from Venus; and found its distance to be about 10°2'·0. Finding Venus very distinct, and consequently the air very clear, I put a magnifying power of 240 times, and, to my great surprise, I found this star put on the same phase with Venus. Its diameter seemed to be about a third, or somewhat less, of the diameter of Venus; the light was not so bright or vivid, but exceeding sharp and well defined. A line, passing through the centre of Venus and it, made an angle with the equator of about 18 or 20 degrees.

'I saw it for the space of an hour several times that morning; but the light of the Sun increasing, I lost it about a quarter of an hour after eight. I have looked for it every clear morning since, but never had the good fortune to see it again.

'Cassini, in his *Astronomy*, mentions another such observation.

'I likewise observed two darkish spots upon the body of Venus; for the air was exceeding clear and serene.'

On 20 May 1759, at 8h 44m, Mayer reported the satellite.[3]

'I saw above Venus a little globe of inferior brightness, about 1 to 1½ diameters of Venus from herself.... The observation was made with a Gregorian telescope of 30 inches focus. It continued for half an hour, and the position of this little globe with regard to Venus remained the same, although the direction of the telescope had been changed.'

Two years later, further observations seemed to confirm the satellite's real existence. A German astronomer, A. Scheuten, reported that during the transit of 1761 he had detected a small black spot following Venus across the Sun's disk, remaining visible even when Venus itself had passed off the Sun.[4] And then Montaigne of Limoges, using a 23-cm refractor, produced a series of observations which sounded most convincing.[5]

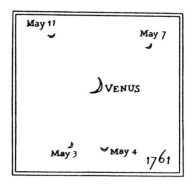

Fig. 11
Position of Venus'
satellite in 1761,
according to Montaigne.

Montaigne first reported the satellite on 3 May 1761, and described it as a little crescent-shaped body about 22 minutes of arc away from Venus. As usual, it showed the same phase as the planet itself, and had one-quarter the diameter of Venus. Montaigne repeated the observation several times during the night, and on 4, 7 and 11 May (the intervening nights were cloudy) he saw the companion again, differently placed but still showing the same phase. Montaigne, who had hitherto been decidedly sceptical about the existence of the satellite, was convinced. He stated that he had taken every possible precaution against optical illusion, and that he had seen the companion even when Venus itself was put outside the field of view.

All this seemed definite enough. In a memoir read to the French Académie des Sciences, Baudouin announced:[6]

'The year 1761 will be celebrated in astronomy, in consequence of the discovery that was made on 3 May of a satellite circling round Venus. We owe it to M. Montaigne, member of the Society of Limoges.... We learn that the new star has a diameter one-quarter that of Venus, is distant from Venus almost as far as the Moon from our Earth, has a period of 9 days 7 hours.'

In 1773 the German astronomer J. Lambert calculated an orbit which gave the mean distance from Venus as about 417,000 kilometres, with a period of 11 days 5 hours, an orbital inclination of 64°, and a orbital eccentricity of 0·195.[7] Frederick the Great of Prussia proposed to name the satellite 'D'Alembert', in honour of his old friend Jean D'Alembert, but the prudent mathematician declined the honour with thanks.

Further observations were made on 3 and 4 March 1764 by Rœdkiær, from Copenhagen;[8] on 10 and 11 March by Horrebow, also at Copenhagen;[8] and on 28 and 29 March by Montbaron at Auxerre, who knew nothing about the Danish work. And henceforth, the satellite disappears from the observation books. Schröter could not find it, though he made a special search; neither could Herschel; and neither could Gruithuisen, who carried out a long series of observations. Satellites do not 'softly and silently vanish away', like the hunters of the Snark, and there is no escape from the conclusion that the satellite of Venus never existed at all.

Maximilian Hell discussed the matter in 1766, and pronounced in favour of optical illusion;[9] this was also the view expressed by Boscovich in the following year.[10] Later, Von Ende wondered whether an asteroid might be responsible,[11] a view revived by Bertrand in 1875.[12] At any rate, old myths die hard, and the ghost moon still had its supporters well into the second part of the nineteenth century. Admiral Smyth, author of the famous *Cycle of Celestial Objects*, believed in it, and claimed that 'the satellite is perhaps extremely minute, while some parts of its body may be less capable of reflecting light than others'.[13] This idea was developed in 1875 by F. Schorr, who went so far as to write a small book about it.[14] Schorr revised Lambert's original period to 12d 4h 6m, and argued that the many failures to see the satellite were due to the fact that it varied in brightness, and was normally too faint to be visible. The theory seemed improbable even at the time, and in any case it was necessary to define just what was meant by 'the satellite'. Whereas Cassini, Montaigne and others had described it as being a quarter the size of Venus itself, Rœdkiær and Horrebow in Copenhagen had seen it as a star-like point. Obviously something was wrong in the State of Denmark!

Another theory was produced in 1884 by J. Houzeau.[15] Originally Houzeau had believed in the existence of the satellite, but then, following a full analysis, rejected it 'first on account of the impossibility of properly representing the observed positions by an orbit described around Venus, and further because the mass of the planet deduced from the least defective attempts would be seven times the real amount'. On the other hand, Houzeau was reluctant to reject the actual observations, and he suggested that they were due not to a Cytherean satellite, but to a separate planet, moving in an orbit slightly outside that of Venus and with a sidereal period of 283 days. He even suggested a name for it: Neith.

The problem was more or less cleared up in 1887 by Paul Stroobant of Brussels, who published an elaborate memoir[16] in which he reprinted all the observations (thirty-three of them, made by fifteen different astronomers) and subjected them to a critical analysis. Some could be rejected outright, while others, such as Montaigne's, had to be put down to ghost reflections. Yet others could be attributed to faint stars. Horrebow, for instance, may have seen the fifth-magnitude star Theta Libræ, while there is a chance that what Rœdkiær saw was the then unknown planet Uranus. Lambert's orbit, too, failed to survive the test of rigorous analysis, since it required the mass of Venus to be ten times greater than it really is.

It would be easy to mistake a star (or even Uranus) for a satellite unless it were followed for a sufficient number of nights for its apparent motion to show it up in its true guise; and Venus is so brilliant that it is quite liable to give rise to ghost images when seen against a reasonably dark background. (Per Wargentin, a famous Swedish astronomer of the eighteenth century, commented that one of his telescopes never failed to show companions to Venus or any other brilliant body.) Admittedly it is strange that observers so skilled as Cassini and Short fell into an elementary trap, but since the satellite is undoubtedly non-existent there is no alternative.

Before leaving the subject, one more observation must be mentioned. On 13 August 1892, E. E. Barnard, using the 91-cm Lick refractor, recorded a seventh-magnitude star-like object in the same field with Venus. The observation was made only half an hour before sunrise, and there is little chance of an optical ghost here—yet Barnard was able to measure the position, and it does not agree with that of any known star.[17] It is worth noting that some time earlier Barnard had made a special search for a satellite of Venus, and had satisfied himself that it did not exist. J. Ashbrook has made the reasonable suggestion that Barnard recorded a nova, or 'new star', which by bad luck was not seen by anyone else.[18]

Though Venus still presents us with plenty of problems, this one, at least, must be regarded as finally settled. Had there been a satellite of appreciable size, it could not have escaped detection in the Space Age. 'Neith' does not exist, and has never done so; Venus like Mercury, is a solitary wanderer in space.

9 Transits and Occultations

As the Earth's distance from the Sun is greater than that of Venus, there must be occasions when the three bodies move into a direct line, with Venus in the middle. For obvious reasons, this can occur only at inferior conjunctions. Venus may then be seen with the naked eye as a small dark disk silhouetted against the bright background of the Sun's face. Such a phenomenon is known as a transit of Venus.

Were the orbits of the two planets in the same plane, a transit would occur at every inferior conjunction, but this is not the case. The orbit of Venus is inclined at an angle of 3·4 degrees, and transits are rare. At the present epoch they are seen in pairs, the components of a pair being separated by eight years—after which no more transits are seen for over a century. Thus there were transits in 1631, 1639, 1761, 1769, 1874 and 1882; the next will be on 8 June 2004, 5–6 June 2012, 11 December 2117, 8 December 2125, 11 June 2247, 9 June 2255, 12–13 December 2360 and 10 December 2368. Calculations for transits after 2012 are of academic interest only so far as we are concerned. In a paper dealing with transit calculations, J. Meeus[1] points out that while the 2004 transit will be wholly visible from London, only the end of the 2012 transit will be seen from there, since the Sun rises while the transit is in progress.[1]

At other inferior conjunctions Venus passes above or below the Sun in the sky, and telescopically it may be followed without a break. In 1950, for example, photographs taken with the 41-cm telescope on Mule Peak, New Mexico, showed Venus when only $7\frac{1}{2}$ degrees from the centre of the Sun's disk.[2]

Obviously, Mercury and Venus are the only planets which can appear in transit (not counting various small asteroids, which are much too small to be seen when passing across the solar face). In 1627 the great mathematician Johannes Kepler finished what was destined to be his last work, a set of new and more accurate tables of planetary movements which he named the Rudolphine Tables in honour of his old benefactor Rudolph II, and he was able to show that both Mercury and Venus would transit the Sun during the year 1631—Mercury on 7 November and Venus on 6 December.[3] By 1631 Kepler was dead, but the transit of Mercury was successfully observed by the French astronomer Pierre Gassendi.[4]

Encouraged by this success, Gassendi naturally expected to be equally fortunate with the transit of Venus, since Venus is not only closer to us than Mercury but is also much larger. He left nothing to chance. Fearful that Kepler's prediction might be in error, he began watching the Sun on 4 December—between breaks in the clouds—and went on throughout 6 and 7 December. To his surprise, and disappointment, he saw nothing. The reason is now known: the transit did indeed occur as Kepler had predicted, but it took place during the northern night of 6–7 December, when the Sun was below the horizon from France.

Kepler had predicted no more transits until 1761, but fresh calculations were made by a young English amateur, the Rev. Jeremiah Horrocks, curate of Hoole in Lancashire, showing that a transit would occur on 24 November 1639 (old style; the new style date is 4 December). Horrocks' calculations were not finished until shortly before the transit was due, and he had time only to inform his brother Jonas, near Liverpool, and his friend William Crabtree, who lived near Manchester.

Horrocks began to watch the Sun on 23 November. On the following day he began observing at sunrise, but was called away to his clerical duties until about ten o'clock—unfortunately it happened to be a Sunday—and he was not able to return to his telescope until 3.15. Then, in his own words:[5]

'At this time an opening in the clouds, which rendered the Sun distinctly visible, seemed as if Divine Providence encouraged my aspirations; when, O most gratifying spectacle! the object of so many earnest wishes, I perceived a new spot of unusual magnitude, and of a perfectly round form, that had just wholly entered upon the left limb of the Sun, so that the margin of the Sun and spot coincided with each other, forming the angle of contact.'

He followed the planet until sunset, an hour and a half later, and was able to make some useful measurements. Crabtree had bad luck with the weather, but he did manage to see Venus just before sunset, when the clouds broke up for a few seconds.

This was the first observation of a transit of Venus; it is true that suggestions had been made that Arab astronomers had witnessed one as long ago as the year 839, but it seems that what was seen was only a sunspot.[6] Horrocks' prediction was undoubtedly a brilliant piece of work.[7] Had he lived, he would probably have ranked with the greatest astronomers of his time, but unfortunately he died in 1641 at the early age of twenty-two.

Before the 1761 transit took place, Edmond Halley, following up an earlier suggestion by James Gregory, realized that these transits of Venus might be used to measure the distance of the Sun from the Earth.[8] Since our world is a globe, and not a point, the position of Venus against the Sun, and hence the times of entry on to and departure from the solar disk, will not be the same for different observing stations; and by careful timing, information can be gained which will lead to the determination of the 'astronomical unit', the distance between the Earth and the Sun. As the whole method is now obsolete there is no point in describing it further, but at all events the transits of 1761 and 1769 were carefully studied from all over the world. Results from the 1761 transit were unsatisfactory, but those for 1769 were rather better, and the Sun's distance was calculated as being about 153,000,000 kilometres, which is of the right order even though slightly too great.[9]

It was in 1761 that Mikhail Lomonosov made the observations which led him to infer the presence of an atmosphere surrounding Venus. And in 1769 David Rittenhouse, in America, saw as illuminated that part of the edge of the planet which was off the solar disk, so that the whole outline of Venus was visible. This could be (and was) due to the Cytherean atmosphere. Probably Rittenhouse was unaware of Lomonosov's earlier observations, but the existence of an atmosphere round Venus seemed highly probable— though it was not really well established until the work by Schröter some time later.

In 1761 Western Europe was useless as an observing site, as only the end of the transit was visible with a very low Sun. Therefore, various expeditions were dispatched. Nevil Maskelyne, later Astronomer Royal, went to St Helena with his colleague Robert Waddington; Charles Mason and Jeremiah Dixon observed from the Cape, and so on.[10] Altogether there were sixty-two observing stations, and 120 sets of timings were made. Unfortunately, the accuracy of the observations was ruined by an effect known as the

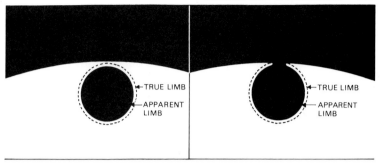

Fig. 12 The Black Drop.

Black Drop. As Venus passes on to the Sun, it seems to draw a strip of blackness after it, and when this disappears the transit has already begun, so that no exact timings can be made. The 1761 results gave measurements of the solar parallax ranging between 8″.28 and 10″.60, which would give values for the astronomical unit of between 160,000,000 kilometres and only 129,000,000 kilometres.

However, there was another chance in 1769, and despite the Black Drop astronomers did not propose to miss it. Nevil Maskelyne was particularly enthusiastic, and at a special Royal Society meeting on 17 November 1767 he was one of four Fellows making special recommendations (the others were John Bevis, James Short and James Ferguson). Eventually, parties were sent to Hudson's Bay (William Wales and James Dymond), North Cape (William Bayly), Hammerfest (Jeremiah Dixon) and north-west Ireland (Mason). Finally there was the South Seas expedition, which brings us to the story of Captain James Cook.

A ship was fitted out expressly for the purpose, with the full agreement of King George III, who was an astronomical enthusiast and actually watched the 1769 transit from the newly-founded observatory at Kew.[11] The chosen observer was Alexander Dalrymple, but when he was refused command of the ship he withdrew in a huff, and in his place the Admiralty called in James Cook. The senior astronomer was Charles Green, who had been an assistant at Greenwich Observatory, and was well known to Maskelyne. After some deliberation the Admiralty decided that the island of Tahiti would be the best site, and on 26 August 1768 the *Endeavour* sailed from Plymouth. By 10 April 1769 the ship was standing off Tahiti, or, to give it the natives' own name, Otaheite. (Europeans had first heard of it in 1607, when it had been discovered by Pedro Quiros; Captain Samuel Wallis rediscovered it in 1767 and named it King George's Island, which is what Cook called it.)

A temporary observatory was set up, and there were no mishaps, apart from the fact that an essential quadrant was stolen by one of the local inhabitants, and was recovered with some difficulty. On the day of the transit the weather was perfect, and observations were made by Cook, Green and Daniel Solander. Cook's own account reads as follows:[12]

'The first appearance of Venus on the Sun was certainly only the penumbra, and the contact of the limbs did not happen till several seconds after . . . it appeared to be very difficult to judge precisely of the times that the internal contacts of the body of Venus happened, by reason of the darkness of the penumbra at the Sun's limb, it being there nearly, if not quite, as dark as the planet. At this time a faint light, much weaker than the rest of the penumbra, appeared to converge toward the point of contact, but did not quite reach it, see Fig. 2. This was seen by myself and the other two observers, and was of great assistance to us in judging of the time of internal contacts of the dark body of Venus, with the Sun's limb. . . . I judged that the penumbra was in contact with the Sun's limb 10″ sooner than the time set down above; in like manner at the egress the thread of light was wholly broke by the penumbra. . . . The breadth of the penumbra appeared to me to be nearly equal to $\frac{1}{8}$ of Venus' semi-diameter.'

Cook gave the apparent diameter of Venus as 56 seconds of arc (the maximum apparent diameter as seen from the Earth is in fact about 66 seconds).

The *Endeavour* left Tahiti on 13 July, and continued on the voyage of exploration which is part of our history. The real tragedy was that when the ship was near Java, Charles Green died.

The transit was observed from eighty stations in all, and there were 150 sets of measurements. The results were, on the whole, better than those of 1761. They were carefully analysed by the Finnish mathematician Anders Lexell, who gave a value for the solar parallax of 8″·63, corresponding to an astronomical unit of 153,000,000 kilometres, while Simon Newcomb later revised this to 8″·79, which is very close to the truth. Yet it was painfully clear that because of the Black Drop, the whole method was much less satisfactory than had been hoped.

Historical references to the 1761 and 1769 transits would be incomplete without mentioning the incredible series of misfortunes suffered by a French astronomer, Guillaume Legentil.[13] Knowing that the 1761 transit would be favourably visible from India, he set out in the previous year. Originally he meant to go to Rodriguez, but decided to change his observing site to Pondicherry, and he made his way there on a French frigate. Unfortunately a war between England and France was in progress, and about this time Pondicherry fell to the English, so that Legentil had to turn back. Before he could reach land the transit was over, and all he could do was to make some rough notes from the ship, using improvised equipment. Rather than risk a second delay he elected to wait in India for the next eight years, and observe the 1769 transit instead. Again he altered his observing site: again he was unlucky. The transit occurred on 3 June 1769. June 1 and 2 were glorious days, but the 3rd itself was hopelessly cloudy, and Legentil saw nothing at all, though his companions who had remained behind at the original site had a perfect view. It was rather too long to wait for the next transit (that of 1874), and accordingly Legentil packed up what belongings he could, and set off for home. Twice he was shipwrecked, and eventually reached France, after a total absence of eleven years—to find that he had been presumed dead, and that his heirs were preparing to distribute his property. . . .

Despite the Black Drop, new preparations were made for the next pair of transits, those of 1874 and 1882. Again there were many observing stations and many sets of measurements: again the results were disappointing.[14] Since there are now much more accurate methods of measuring the length of the astronomical unit, the next transits, those of 2004 and 2012, will be regarded as of no more than academic interest, but at least they will be fascinating to watch. Remember, the last transit occurred almost a century ago, so that there can be nobody now living who remembers a transit of Venus.

Transits of Venus could, of course, be observed from other planets—if we could get there! On 16 July 1910, for instance, a transit would have been seen by an observer on Saturn lasting for over eight hours.[15] To our hypothetical Saturnian, however, Venus would appear very small, with an apparent diameter of only 1·97 seconds of arc.

As the Moon moves across the sky, it may pass in front of, and occult, a star. When this happens, the star remains visible right up to the lunar limb, and then snaps out abruptly, owing to the lack of atmosphere round the Moon's edge. When a planet is occulted, it takes some seconds to be covered by the oncoming limb of the Moon. Venus is occulted now and then; a particularly beautiful photograph of the planet almost on the Moon's limb was taken from Tokyo, in 1934, by K. Suguki.[16]

Occasionally, one planet may be occulted by another. This happened on 3 October 1590, when an occultation of Mars by Venus is said to have been watched by Michael Möstlin, Professor of Mathematics at Heidelberg;[17] and again on 17 May 1737, when Mercury was occulted by Venus. The latter occultation was seen by John Bevis from Greenwich, though his view was interrupted by clouds.[18] (A writer named Simonelli infers that it was also seen by J. J. Cassini from Paris,[19] but this is a mistake; Cassini says distinctly that he failed to do so,[20] since Mercury was still well clear of Venus when both planets were lost in the horizon mist.[21]) On 21 July 1859 Venus and Jupiter were so close together in the sky that they could not be separated without a telescope, though there was no actual occultation.

It may be of interest to list the planetary conjunctions involving Venus between 1800 and 2100. The following list gives the conjunctions in which the separation between the centres of the planets is less than 60 seconds of arc, and the distance from the Sun is more than 10 degrees:

Planets	Date	GMT	Separation, "	Elongation, °
Venus–Neptune	8 Jan 1813	08.36	60	35 W
Venus–Jupiter	3 Jan 1818	21.52	12 occ.	16 W
Venus–Saturn	8 Feb 1847	01.15	43	13 E
Venus–Jupiter	21 July 1859	03.47	32	19 W
Venus–Jupiter	6 Feb 1892	10.16	40	34 E
Venus–Neptune	27 Apr 2022	19.21	26	43 W
Venus–Neptune	15 Feb 2023	12.35	42	28 E
Venus–Uranus	20 Jan 2077	20.01	43	11 W

Venus will occult Jupiter on 22 November 2065, at 12.46; the separation between the centres will be 14 seconds of arc, but the pair will be only 8° west of the Sun.

When Venus occults a star the observations are interesting, though obviously of less value than they used to be before we had any reliable knowledge about the depth of the Cytherean atmosphere. The star fades appreciably before being occulted by the solid body of the planet. Unfortunately, the phenomena are difficult to observe. For instance, the Japanese astronomer I. Yamamoto relates how he made a special journey to the Kurasiki Observatory, near Okayama, to observe the occultation of a sixth-magnitude star on 15 December 1927, but failed to detect the star until two minutes after it had reappeared from behind Venus.[22]

Better fortune was experienced on 26 July 1910 by E. M. Antoniadi, F. Baldet and F. Quénisset at the Flammarion Observatory, near Juvisy in France. Using telescopes up to 23-cm aperture, they were able to watch the occultation of the star Eta Geminorum, which is somewhat variable, but is never as bright as the third magnitude or as faint as the fourth. Their report reads:[23]

'The emersion occurred under favourable conditions, so that we were able to confirm, independently, and very clearly, that Eta Geminorum (then of magnitude $3\frac{1}{2}$) did not reappear suddenly, as with lunar occultations. In fact, there was at first a barely perceptible luminescence; then the very faint star seemed to detach itself from the dark edge of the planet. It rapidly increased in brilliancy, and in $1\frac{1}{2}$ to 2 seconds after its first appearance had regained its initial brightness. Besides this increase in luminosity at the moment of emersion, we noticed that Eta Geminorum continued to gain slowly and slightly in intensity according to its distance from Venus. . . . There was no appreciable change in the colour of the star. . . . The hypothesis which appears to us the most probable for explaining the variation in brightness is that the light of the star was absorbed in traversing the atmosphere of Venus. In our observation, this variation, which lasted from $1\frac{1}{2}$ to 2 seconds, corresponds to a motion of the planet from $0''\cdot8$ to $1''\cdot1$. We hence deduce the height of the atmosphere of Venus which produced this absorption to be from 50 to 70 miles' [i.e. from about 80 to about 113 kilometres].

This sort of behaviour was confirmed in 1918, when W. W. Campbell, using the 91-cm Lick refractor, watched the occultation of the star 7 Aquarii, on 2 March. According to Thiele, 'observation shows that the fading of the light of 7 Aquarii was appreciable for only a few minutes before the star's disappearance [perhaps due to] some opaque and changing hindrance in the light, like high clouds in the atmosphere of Venus'.[24]

On 19 March 1948 Venus occulted the star 36 Arietis, and observations carried out by six members of the Association of Lunar and Planetary Observers, in the United States, gave an atmospheric height of just over 110 kilometres.[25]

On 7 July 1959 Venus occulted the first-magnitude star Regulus. The phenomenon occurred in broad daylight, at 14.28 GMT, and one of the present writers (Moore) was fortunate enough to observer it, together with Henry Brinton, at Selsey in Sussex, using Brinton's 30-cm reflector. The fading before occultation was very marked. Many other observations were made, and were analysed by G. de Vaucouleurs and D. H. Menzel;[26] the results for the extent of Venus' atmosphere turned out to be at least reasonably accurate. Regulus is one of the few really bright stars which may be occulted by Venus,[27] but the next occasion will not be for more than half a century. In fact, close conjunctions of Venus with stars of magnitude $2\frac{1}{2}$ or brighter are very rare. The only cases during the period from 1800 to 2100 are as in the table below.

Planets	Date	Time, GMT	Distance, "	Elongation, °
Venus–Regulus	29 Sept 1817	03.24	25	38 W
Venus–Regulus	7 July 1959	14.28	4 occ.	45 E
Venus–Sigma Sagittarii	17 Nov 1981		10 occ.	47 E
Venus–Regulus	1 Oct 2044	22.02	4 occ.	39 W

It cannot be said that in the light of modern research, occultations by Venus have retained their previous importance; but they are always worth watching—and it is fascinating to see the star flicker and fade before being finally hidden as the solid body of Venus sweeps across it.

Summary

This survey has taken us up to the start of the Space Age, and it is interesting to look back at the state of our knowledge (or lack of it) less than twenty years ago.

It had been established that the main constituent of the atmosphere was carbon dioxide, but there was no positive information about the state of affairs well beneath the cloud-tops. Presumably the surface temperature was high, but estimates showed a wide range, and so for that matter did the measurements of the upper clouds themselves. The length of the rotation period was most uncertain; G. P. Kuiper's estimate of 'a few weeks' was regarded as the most likely value, but there were still some astronomers who preferred a period in the region of 24 hours, and in France the 224·7-day synchronous period remained popular. Efforts to map permanent surface features had led to no really useful results. As for the surface—well, it could be either an arid, fiercely hot dust-bowl, or else covered with oceans. If the latter, then there was no reason why primitive life-forms should not have appeared, just as they did in the warm pre-Cambrian seas of Earth. Venus, it was said, might be a world where life was just starting to develop.

Then, on 26 August 1962, Mariner 2 was sent on its way. The new era in Man's exploration of Venus had begun.

Part 2   Venus in the Space Age

10 Rockets to Venus

The Space Age began on 4 October 1957. Russia's Sputnik 1 soared aloft, and entered a closed orbit round the Earth. It was football-sized, and carried little apart from a radio transmitter which sent back the never-to-be-forgotten 'Bleep! bleep!' signals which caused such intense interest (and, frankly, surprise) over much of the world. Two years later Luna 3 began its journey round the Moon, obtaining the first pictures of the averted regions which are never visible from Earth because they are always turned away from us. And on 12 February 1961, the Soviet scientists made their first launching towards Venus.

Venera 1 was, therefore, the first interplanetary probe in history. It was not successful; contact with it was lost when it had receded to 7,500,000 kilometres from Earth, and was never regained, so that nobody will ever know what happened to it (it may well have passed within about a hundred thousand kilometres of Venus in May 1961, and presumably it is still in a solar orbit). In July 1962 the Americans made their first attempt, with Mariner 1, but the result was disastrous; Mariner 1 merely plunged into the sea, apparently because somebody had forgotten to feed a minus sign into a computer (a slight mistake which cost approximately £4,280,000). But then came the triumphant Mariner 2, and the era of direct planetary exploration had well and truly begun.

Sending a probe to another planet is not quite so straightforward as might be expected. It would be convenient to wait until the Earth and the planet are at their nearest (approximately 39,000,000 kilometres in the case of Venus) and then fire a rocket straight across the gap, but there are any number of reasons why this cannot be done. For instance, it would involve using more propellant than any rocket could possibly carry. It is essential to make use of the Sun's gravitational pull, so that the probe can 'coast', unpowered, for most of the journey. Mariner 2 was a case in point. It was launched from Cape Canaveral on 27 August 1962, carried in an

Atlas rocket which moved more or less vertically upwards and then headed off in the general direction of South Africa. The second stage, an Agena rocket, then took over, and the Agena-Mariner combination entered what is termed a parking orbit, at a mean height of 185 kilometres above ground level and moving at 29,000 km/hour. As it reached the African coast, the Agena fired once more, and the total velocity rose to 41,034 km/hour, which is more than the escape velocity for that altitude.

For a Venus probe, the direction of 'escape' has to be opposite to that in which the Earth is moving round the Sun. Therefore, as the Mariner moved out, it was slowed down by the Earth's gravitational pull, until at a distance of 965,000 kilometres the velocity round the Sun had been reduced to 11,060 kilometres less than that of the Earth. It started to swing inwards towards the Sun, gathering velocity as it went. Meanwhile the Agena rocket had been separated from the Mariner, rotated through a wide angle and fired again, so that it moved away in a completely different orbit; its work had been done.

As can be seen from the diagram, (Fig. 15) Mariner 2 then followed a transfer orbit. After one mid-course correction (on 4 September) it swung inwards until it reached the orbit of Venus on 14 December. It was moving considerably faster than Venus, but it passed within 35,000 kilometres of the planet, and sent back invaluable data before it moved out of range and entered a never-ending orbit round the Sun. Contact with it was finally lost on 4 January 1963, when it was some 87,000,000 kilometres from the Earth.

During its period of useful contact with Venus, Mariner 2 revolutionized many of our cherished ideas about the planet. The Whipple–Menzel marine theory was killed at once; the surface was far too hot for water to exist in the liquid form, even under the very high atmospheric pressure. A very long rotation period, as indicated by Earth-based radar, was confirmed, and there appeared to be no detectable magnetic field. In fact, Venus was not a world of the type expected by many astronomers.

Before going any further, it may be useful to list all the Venus probes launched between 1961 and 1980. The Mariners and Pioneers are American; the rest, Russian.

Obviously, these probes were of different types. Few data were released about Zond 1, though it seems to have been aimed at or near Venus. Mariner 10 was essentially a Mercury probe; it

9   The 210 m Goldstone receiver situated in the Mojave Desert, California.

Name	Launch	Arrival	Closest approach, km	Results
Venera 1	12 Feb 1961	19 May 1961	100,000?	Contact lost at 8,500,000 km
Mariner 1	22 July 1962	—	—	Total failure
Mariner 2	26 Aug 1962	14 Dec 1962	35,000	Fly-by. Data transmitted.
Zond 1	2 Apr 1964	?	?	Contact lost in a few weeks.
Venera 2	12 Nov 1965	27 Feb 1966	24,000	In solar orbit. No Venus data received.
Venera 3	16 Nov 1965	1 Mar 1966	Landed	Crushed during descent. No Venus data received.
Venera 4	12 June 1967	19 Oct 1967	Landed	Data sent back during 94-min descent.
Mariner 5	14 June 1967	19 Oct 1967	4,000	Fly-by. Data transmitted.
Venera 5	5 Jan 1969	16 May 1969	Landed	Data transmitted during descent.
Venera 6	10 Jan 1969	17 May 1969	Landed	Data transmitted during descent.
Venera 7	18 Aug 1970	15 Dec 1970	Landed	Transmitted for 23 min after landing.
Venera 8	26 Mar 1972	22 July 1972	Landed	Transmitted for 50 min after landing.
Mariner 10	3 Nov 1973	5 Feb 1974	5,800	Pictures of upper clouds; data transmitted. Went on to Mercury.
Venera 9	8 June 1975	21 Oct 1975	Landed	Transmitted for 53 min after landing. One picture received.
Venera 10	14 June 1975	25 Oct 1975	Landed	Transmitted for 65 min after landing. One picture received.
Pioneer Venus 1	20 May 1978	4 Dec 1978	145	In orbit. Data sent back.
Pioneer Venus 2	8 Aug 1978	9 Dec 1978	Landed	Multiprobe. 5 probes landed. Data sent back.
Venera 11	9 Sept 1978	21 Dec 1978	Landed	Transmitted for 60 min after landing. Data received from orbiter.
Venera 12	14 Sept 1978	25 Dec 1978	Landed	Transmitted for 60 min after landing. Data received from orbiter.

Name	Launch	Arrival	Closest approach, km	Results
Venera 13	30 Oct 1981	1 March 1982	Landed	Data received from the main station in solar orbit. Transmitted for 127 minutes.
Venera 14	3 Nov 1981	5 March 1982	Landed	Landing position lat. S. 13° 151' long. 310° 9'. Soil analysis carried out.

by-passed Venus *en route* and sent back some excellent pictures of the cloud tops, as well as other information, but Mercury was its main objective, and in fact it has provided us with our only detailed information to date about the surface of that planet.[1] Generally speaking, the first American vehicles were of the fly-by variety, while the main Russian objective was to soft-land. What the Soviet researchers did not know, and could not be expected to know, was that the great pressures in the Cytherean atmosphere were enough to put any but the sturdiest vehicle out of action well before a landing could be made—simply by crushing it. This explains the apparently contradictary early results. It was also necessary to chill a probe before bringing it down through the extremely hot lower atmosphere. All in all, the surprising thing is not that the Russians had initial failures, but that they succeeded as soon as they did. Up to 1982, the only pictures received direct from the Cytherean surface have been those from Veneras 9, 10, 13 and 14. This makes it all the more remarkable that the Russians have had such scant success with Mars, which should be a much easier target in every way.

The first American landings on Venus were delayed until December 1978. Even then there was no real effort to allow any of the probes to survive the shock of impact (though in fact one of them did so for a brief period). Meanwhile, it is fair to say that since the Mariner 2 flight of 1962, we have learned more about Venus than had been possible throughout the whole of human history.

During the past decade, the Soviet space programme to Venus has been primarily associated with entry probe missions to penetrate the atmosphere and directly examine the surface properties.[2] Venera 4 was the first successful mission to enter the atmosphere; after aerodynamic braking, it descended by parachute from an altitude of 50 km, returning information on the atmospheric properties until transmissions ceased at about 29 km. An 'interplanetary bus' subsequently burned up in the atmosphere. Venera 4 found out a great deal about the atmosphere, but several of its instruments went 'off scale'; however, it was established that the chief atmospheric constituent was carbon dioxide, with suggestions of traces of water vapour. At this stage nitrogen was assumed to be present.

The data sent back from Venera 4 were confirmed and extended by the subsequent Soviet missions to the planet, which were all entry probes in the Venera series.[3] Veneras 5 and 6 entered the atmosphere on the dark side of the planet within 24 hours of each other at points separated by about 300 km. Both landers carried similar experiments to those of Venera 4, with an increased range and sensitivity, and measured atmospheric parameters throughout a parachute descent from 55 to 20 km before contact was lost. As a result, the surface properties of temperature and pressure were obtained by extrapolating the observations.[4,5]

Venera 7 was modified to withstand the most extreme surface temperatures and pressures that had been estimated from the results of previous missions. Although instrument ranges were extended to 180 atmospheres and 800°K, a telemetry failure meant that only temperature data were obtained during the descent of the entry probe. However, Doppler tracking revealed a sudden change in the velocity of the probe, followed by a period of twenty-three minutes in which the temperature and velocity readings remained constant. This was interpreted as a successful landing.[2,6]

Venera 8 met with much greater success. After a 55-minute descent, during which the atmospheric composition was regularly monitored, the entry probe landed on the illuminated side of the planet about 600 km from the morning terminator. These observations showed that the temperature profile on the day-side was essentially indistinguishable from that of the night-side measurements obtained from the earlier entry probes. The Doppler tracking of the space-craft indicated that the surface winds were less than 1 m/second, increasing to about 100 m/second at an altitude of 60 km.[2,7] It appeared that the downward solar flux measurements had a discontinuity at 32 km altitude, which was suggestive of a change in the cloud structures. Also, only 1 per cent of the incident solar energy actually reached the surface of Venus. This is rather

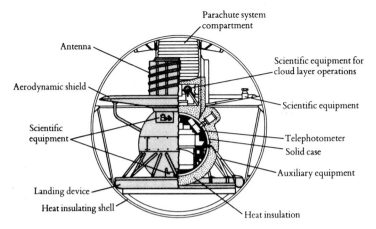

Fig. 13   Structure of the Venera 9 lander.

like being in London on a very overcast winter day. Venera 8 also carried a chemical colour indicator to search for alkaline gases, particularly ammonia, which gave a positive result. This observation has not been confirmed by subsequent, improved instruments carried in later probes.

Veneras 9 and 10 continued the series; they were large vehicles based upon the experience gained from previous missions, but now each system consisted of an entry probe and an orbiter section. The orbiter sections were the first vehicles to be placed in permanent orbit round Venus. The landers entered the atmosphere and, after aerodynamic braking, fell slowly by parachute from 65 to 50 km, to give prolonged coverage of the cloud layer. At 50 km, the parachutes were released, allowing a more rapid descent through the hot lower atmosphere controlled by a braking shield. Both landers struck the surface at about 7 m/second on the illuminated side of the planet, and proceeded to relay photographs and other surface data to the Earth via their parent orbiters until they passed out of range after about one hour.[2,8] The orbiters carried instruments designed to study features of the upper atmosphere and the cloud tops.

Veneras 11 and 12 were essentially similar to the previous probes, carrying both orbiter and lander systems.[2] Their instruments had been considerably improved, but unfortunately the television systems, designed to send back images of the landing area, failed in both space-craft. This in itself underlined the fundamental problems of obtaining pictures of the surface in these intensely hostile conditions.

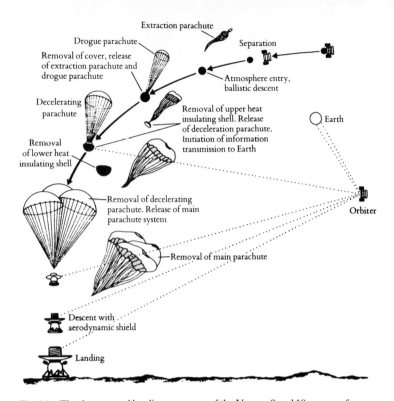

Fig. 14   The descent and landing sequence of the Venera 9 and 10 spacecraft.

Then, in 1982, came the most successful of all the Russian probes: Veneras 13 and 14. Each included a lander, which transmitted its data via the remainder of the vehicle which continued in solar orbit. Colour pictures from the surface were obtained, and soil sampling was undertaken for the first time.

Throughout the period of the active Soviet programme, the United States has launched only two major space missions to Venus. The first was Mariner 10, which really used Venus as a means of deflecting the space-craft on to a rendezvous with the innermost planet, Mercury, which was its prime target. Mariner 10 carried a payload similar to that of its predecessor Mariner 5, designed to study the upper atmosphere of Venus. In addition, it carried a two-channel infra-red radiometer and an imaging system.[9] Some of the most striking results from Mariner 10 were the image mosaics of Venus obtained at ultra-violet wavelengths which provided important information with regard to the patterns of the cloud tops. The rapid 4-day circulation at equatorial latitudes was evident, and the variation in the structure of the winds towards the

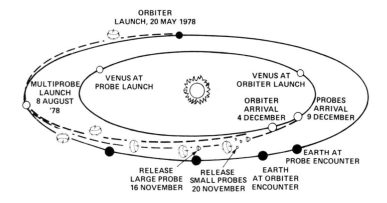

Fig. 15   The flight path of the Pioneer Venus spacecraft.

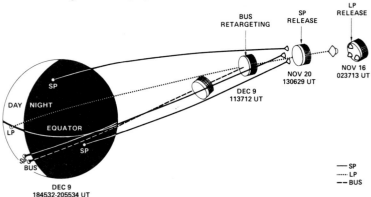

Fig. 16   The separation sequence for the Pioneer Venus multiprobes.

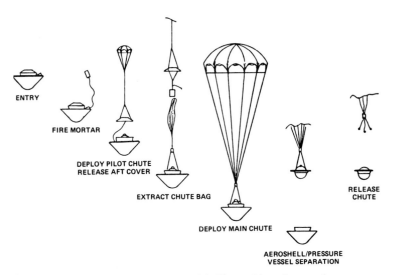

Fig. 17   The descent sequence of the Pioneer Venus large probe.

poles was observed for the first time. The meteorology of the upper Cytherean atmosphere certainly appeared more complex than had been assumed from earlier measurements made with Earth-based telescopes.

The Pioneer Venus mission represents a further step in the programme of the exploration of the Cytherean atmosphere. This mission involved two space-craft: an orbiter, and a multiple entry probe ('multiprobe') which consisted of one large and three small entry probes mounted on a 'bus', which was similar in design to the orbiter. The multiprobe concept had the advantage of being able to carry out simultaneous studies of the atmosphere in various locations, on both the day-side and the night-side.[10] The orbiter reached Venus on 4 December 1978 (see Fig. 20). The orbital characteristics were as follows:

Periapsis	150–200 km
Apoapsis	66,900 km
Eccentricity	0·843
Average period	24.03 h
Inclination to the equator	105·6°
Periapsis latitude	17·0°N
Periapsis longitude (Orbit 5)	170·2°

The five separated space-craft of the multiprobe mission (bus, large sounder probe, small sounder probe, small day probe, small night probe) encountered Venus on 9 December 1979. These space-craft had an unusual approach to the planet (Fig. 16). Some four hours before the encounter on 16 November, the spin-stabilized bus was oriented, and the large (sounder) probe was released towards the planned entry location on Venus. About 4·5 days later, on 20 November, the three small probes (sounder, day, night) were released at a precise spin rate (49·6 rpm) and precise time in the bus spin cycle towards their planned Venus entry locations. All four probes were then silent until 22 minutes before entry on 9 December.

The entry of all the four probes into the atmosphere at about 200 kilometres above the surface of the planet was staggered over a period of about ten minutes, which then permitted the ground stations to receive sequentially the telemetry signals from the individual probes. After entry, atmospheric braking began well above 100 kilometres, reaching a peak deceleration of 500–600 g near 78 km. At 68 km, the main parachute of the large probe opened, and the probe descended slowly through the cloud layer. After 17 minutes, at approximately 47 km, this parachute was jettisoned, allowing the probe to fall aerodynamically until it struck the surface 39 minutes later.

The small probes did not carry parachutes, and fell freely from 60 km to the surface in a similar period. Although none of the probes was designed to operate after impact, the day-probe continued making measurements for 67·37 minutes after arriving on the surface of Venus. In Fig. 18 we show the landing sites of all the Venus spacecraft from the US and Soviet space programmes.

The range of instruments on the probes and the entry bus were designed to investigate the composition and structure of the Cytherean ionosphere, neutral atmosphere and the clouds that obscure the surface. Several of the instruments on the probes had external sensors, and all these failed at altitudes of about 12 km. These failures may well be scientifically very interesting, although the reasons for them are not yet fully understood. Subsequent analyses suggest that the failures may be associated with the external electric discharge phenomena experienced by all the probes.

The nominal 243-day orbiter mission was completed on 4 August 1979. The space-craft was still operating in 1982, and is expected to

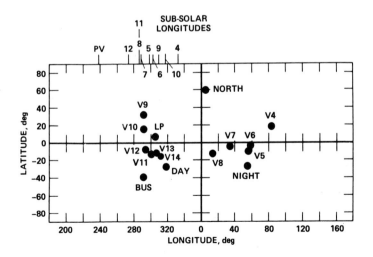

Fig. 18   The co-ordinates of the landing sites of the Venus probes.

Fig. 19 (a–f)   Pioneer Venus vehicles

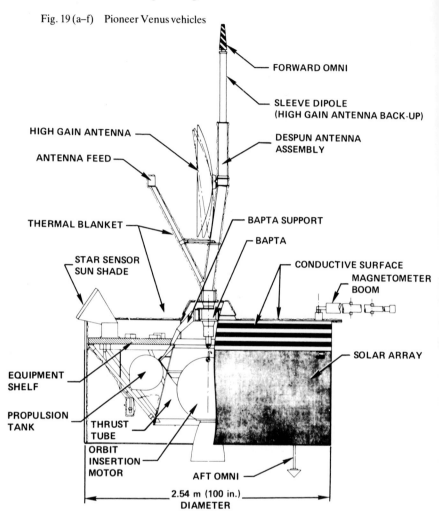

FORWARD OMNI

SLEEVE DIPOLE
(HIGH GAIN ANTENNA BACK-UP)

HIGH GAIN ANTENNA

ANTENNA FEED

DESPUN ANTENNA
ASSEMBLY

THERMAL BLANKET

BAPTA SUPPORT

BAPTA

STAR SENSOR
SUN SHADE

CONDUCTIVE SURFACE
MAGNETOMETER
BOOM

SOLAR ARRAY

EQUIPMENT
SHELF

PROPULSION
TANK

THRUST
TUBE

ORBIT
INSERTION
MOTOR

AFT OMNI

2.54 m (100 in.)
DIAMETER

(a) Orbiter configuration

remain active for several more years yet, continuing the study of the
upper clouds and the surrounding environment. During this time
there will be considerable variations in solar activity, which will be
of great help in interpreting the results.

The Soviet Venera 11 and 12 missions were also carried out
around the same time. They encountered Venus on 21 and 25
December 1978 respectively (see table, page 116). In Fig. 18 we

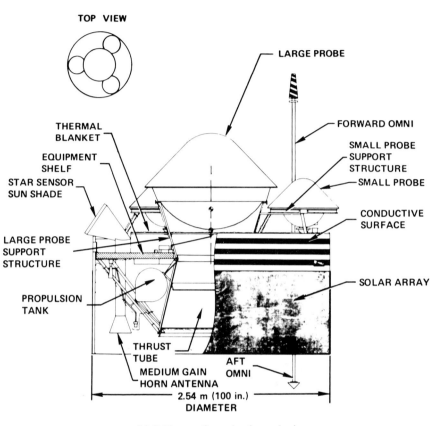

TOP VIEW

LARGE PROBE

THERMAL
BLANKET

FORWARD OMNI

SMALL PROBE
SUPPORT
STRUCTURE

EQUIPMENT
SHELF

STAR SENSOR
SUN SHADE

SMALL PROBE

CONDUCTIVE
SURFACE

LARGE PROBE
SUPPORT
STRUCTURE

PROPULSION
TANK

SOLAR ARRAY

THRUST
TUBE

AFT
OMNI

MEDIUM GAIN
HORN ANTENNA

2.54 m (100 in.)
DIAMETER

(b) Orbiter configuration (near view)

have illustrated the positions in which all these space-craft have
landed on the surface of the planet. Veneras 13 and 14 of 1982 were
even more ambitious, demonstrating in particular that active
vulcanism may be in progress.

Altogether, these missions have given us a new insight into our
understanding of Venus. For the first time we have been able to see
beneath the clouds, and map the surface features from radar
equipment carried in the orbiter of Pioneer Venus. These
observations have provided a whole new concept of our nearest
planetary neighbour.

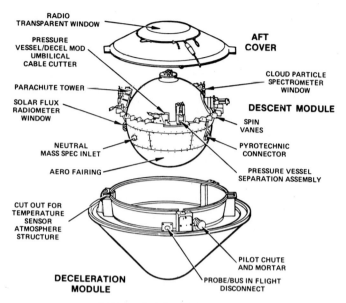

(c) Expanded view of the large probe

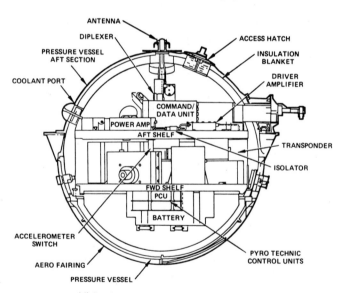

(d) Section of the large probe descent module

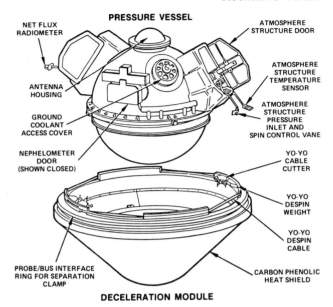

PRESSURE VESSEL

NET FLUX RADIOMETER

ATMOSPHERE STRUCTURE DOOR

ATMOSPHERE STRUCTURE TEMPERATURE SENSOR

ANTENNA HOUSING

ATMOSPHERE STRUCTURE PRESSURE INLET AND SPIN CONTROL VANE

GROUND COOLANT ACCESS COVER

NEPHELOMETER DOOR (SHOWN CLOSED)

YO-YO CABLE CUTTER

YO-YO DESPIN WEIGHT

YO-YO DESPIN CABLE

PROBE/BUS INTERFACE RING FOR SEPARATION CLAMP

CARBON PHENOLIC HEAT SHIELD

DECELERATION MODULE

(e) Expanded view of the small probe

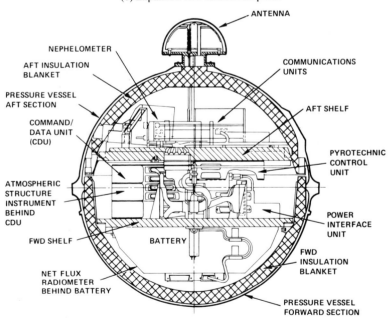

ANTENNA

NEPHELOMETER

AFT INSULATION BLANKET

PRESSURE VESSEL AFT SECTION

COMMAND/ DATA UNIT (CDU)

ATMOSPHERIC STRUCTURE INSTRUMENT BEHIND CDU

FWD SHELF

NET FLUX RADIOMETER BEHIND BATTERY

COMMUNICATIONS UNITS

AFT SHELF

PYROTECHNIC CONTROL UNIT

POWER INTERFACE UNIT

BATTERY

FWD INSULATION BLANKET

PRESSURE VESSEL FORWARD SECTION

(f) Section of the small probe pressure vehicle

11 The Atmosphere of Venus

Composition

Had there been any intelligent life on Venus, it would have had a
tremendous shock in December 1978, when suddenly seven
space-craft landed on the surface of the planet within a few days of
each other. Two were Russian: Veneras 11 and 12. The others were
components of the Pioneer Venus mission.

Each vehicle carried instruments designed to make accurate
measurements of the atmospheric composition at different levels
above the visible clouds. The objective was to provide an
unambiguous picture of the vertical distribution of the constituents,
together with an indication of the possible spatial deviations.

The observations confirmed the previous deduction that carbon
dioxide is the major constituent, with the content of nitrogen,
oxygen, sulphur dioxide and water vapour amounting to a few per
cent. However, the main interest was how the atmosphere of Venus
compared with that of the Earth both in its present condition and in
its evolutionary history. One basic requirement is to measure the
amounts of the inert gases, which are chemically unreactive, and
whose amounts are the same today as they were when the planet was
formed. This cannot be done from observations made from outside
the atmosphere. The only method is to analyse a sample of the
atmosphere itself.

The table below lists the isotopic ratios of some of the substances
found in the atmosphere of Venus.[1] The atmosphere seems to
contain far more of two isotopes of argon (Ar-38 and Ar-36) than
the Earth's atmosphere, but this is not so for Ar-40, which is
produced by the radioactive decay of an isotope of potassium. This
situation is certainly curious. On Earth, about 10 per cent of the
argon produced by radioactive processes is released into the
atmosphere because of the weathering processes of the rocks. On
the Moon, where no weathering occurs, the present rate of emission
of Ar-40 accounts for about 6 per cent of the total production rate,

implying that other mechanisms of release of trapped gases are as important as weathering. Therefore, it would be expected that the amount of Ar-40 in the atmosphere of Venus would be quite large. Perhaps, therefore, the lower abundance actually found there suggests that Venus has less potassium than the Earth, or that the release mechanism is less efficient than with the Earth.

Isotopic ratios of some substances found in
the atmosphere of Venus

Gas	Venus atmosphere Isotopic ratio	Earth atmosphere Isotopic ratio
^{3}He/^{4}He	$<3 \times 10^{-4}$	$1 \cdot 4 \times 10^{-6}$
^{22}Ne/^{20}Ne	$0 \cdot 07 \pm 0 \cdot 02$	$0 \cdot 97$
^{20}Ne/^{36}Ar	$0 \cdot 3 \pm 0 \cdot 2$	$0 \cdot 58$
^{38}Ar/^{36}Ar	$0 \cdot 18 \pm 0 \cdot 02$	$0 \cdot 187$
^{40}Ar/^{36}Ar	$1 \cdot 03 \pm 0 \cdot 04$	296
^{13}C/^{12}C	$\leqslant 1 \cdot 19 \times 10^{-2}$	$1 \cdot 11 \times 10^{-2}$
^{18}O/^{16}O	$(2 \pm 0 \cdot 1) \times 10^{-3}$	$2 \cdot 04 \times 10^{-3}$

A further surprise concerns the distribution of krypton in the inner Solar System. The Cytherean atmosphere has far more krypton than might have been expected. Certainly there is less krypton than in the case of the Sun, but more than in the case of Earth. These variations add further important constraints on the origin of the planets.

It is possible that Venus received a large input of various gases from the Sun in the early history of the Solar System, during a period when the solar wind was much denser than it is today. These gases would have impacted the mass of material condensing to form Venus. The proto-Venus would have blocked off this enriched solar wind, preventing it from blowing out to the then forming Earth and Mars. Therefore, today's Earth and Mars are relatively deficient in the two noble gases krypton and xenon as well as in argon.

Indeed, the large excess of argon in the atmosphere of Venus needs careful explanation. The usually efficient way of trapping the noble gases would suggest that Venus may resemble the Sun more closely than the Earth in this respect. Undoubtedly Venus has three times as much krypton as the Earth, and 700 times more primordial

argon than krypton. This compares with an argon:krypton ratio for
the Earth and Mars of only 30 to 1. The Sun, on the other hand, has
2000 times more argon than krypton. The Pioneer Venus results
certainly suggest that these noble gases may indeed have come from
the Sun, forming a veneer on the dust-grains as they condensed out
of the solar nebula.[2]

Owing to the excessive amount of carbon dioxide in the
Cytherean atmosphere as compared with that of the Earth, the
amount of atmospheric nitrogen (N_2) on Venus is about three times
the terrestrial value. However, since the amount of nitrogen fixed in
the Earth's crust is 2 or 3 atmospheres, the total amount of nitrogen
on the two planets is about the same. For Mars, the nitrogen
abundance is far less.

The other important constituents in Venus' atmosphere are
closely associated with the complex chemistry and cloud formation
processes that are taking place there. The distributions of H_2S,
COS, SO_2 and H_2O will all be associated with the sulphur cycle and
the production of sulphur and sulphuric acid droplets into the
various cloud layers. The actual amount of H_2O in the Cytherean
atmosphere is indeed very small compared with that of the Earth; it
is equivalent to only about 10^{-5} the amount contained in our
oceans. The vastly different conditions on the surfaces of the two
planets are due directly to the modest amount of H_2O produced in
the original atmosphere of Venus.

Atmospheric Structure

The many space-craft sent to Venus in recent years have now given
us detailed information about the structure of the atmosphere. The
Pioneer Venus measurements were started at an altitude of 200
km,[3] and in addition observations were made during the initial part
of the mission from 60 to 140 km altitude[4]—a region not previously
studied in a regular way. Fig. 20 shows a comparison of the
temperature profiles of Venus and the Earth, indicating important
differences between the two atmospheres. In the Earth's
atmosphere there are three clearly defined regions characterized by
the changes of sign in the temperature gradient, but the situation
with Venus is not quite the same.

On the day-side of Venus there is a terrestrial-type thermosphere, with temperatures increasing from about 180°K at 100 km to
about 300°K in the exosphere. The thermosphere does not exist on
the night-side of the planet, where the temperature falls from about

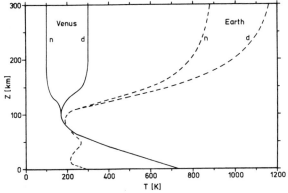

Fig. 20
Comparison of the vertical thermal structures of the atmosphere of Venus and Earth. The day- and night-sides of the profile are shown as (d) and (n) respectively.

180°K at 100 km to as low as 100°K at about 150 km. The transition from day-side to night-side temperatures across the terminator is very abrupt.

Between an altitude of 100 kilometres and the cloud tops at about 70 kilometres, the atmospheric temperature is distinctly variable. Diurnal fluctuations of as much as 25°K have been observed at the 95-km level. It is known that about 90 per cent of the entire atmosphere lies between the surface and a height of 28 km, and at this level the atmosphere is really a massive 'ocean', dense and very sluggish in response to solar heating—which is naturally very feeble at these depths. Fig. 21 shows that there is essentially no difference

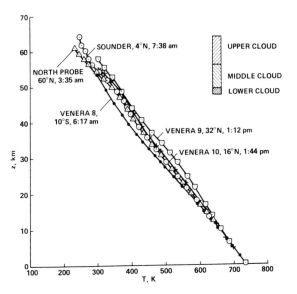

Fig. 21
The temperature structure of the lower Venus atmosphere obtained by comparing the Venera 8, 9, 10 probes with the Pioneer Venus sounder and north probe measurements.

in the atmospheric structure over different parts of Venus. Above 28 km, the differences in the temperature at various locations start to show up. In the upper part of the troposphere, the processes associated with the production and dissipation of the clouds will affect different regions in different ways.

The Greenhouse Effect

The measurements made from a whole range of space instruments have confirmed that the surface temperature of Venus is 737°K. Why is this so much higher than anywhere on the Earth? And why is the Cytherean atmosphere so massive?

These basic questions are connected, since there is as much CO_2 in the atmosphere of Venus as we can find in rocks such as limestone on the surface of the Earth. The basic make-up of the Earth and Venus is, of course, very similar. But Venus moves in an orbit considerably closer to the Sun. The increased heat rapidly warms the surface of the planet, and liberates the CO_2 from it, increasing the opacity of the atmosphere. The solar radiation is able to penetrate the increasing CO_2 atmosphere, so continuing to warm the surface, but the atmosphere is virtually opaque to infra-red radiation emitted by the surface, so that the surface is warmed even further. This process is known as the greenhouse effect;[5] it will continue until the atmosphere and the surface are in a chemical equilibrium. At this stage the surface will be far too hot for water to condense into oceans. We may assume that this is precisely what has happened to Venus, and it also accounts for the very small quantities of H_2O currently found in the atmosphere.

While CO_2 is certainly the major constituent of the Cytherean atmosphere, is this gas alone sufficient to provide the greenhouse effect? Actually, the CO_2 is responsible for about 55 per cent of the trapped heat. A further 25 per cent is due to the presence of traces of water vapour, while SO_2, which constitutes only 0·02 per cent of the atmosphere, traps 5 per cent of the remaining infra-red radiation. The remaining 15 per cent of the greenhouse effect is due to the clouds and hazes which surround the planet.[6]

Could this situation happen on the Earth? Man himself is causing the amount of atmospheric CO_2 to increase by burning fossil fuel (coal) and at the same time reducing the sink for removing CO_2 by world-wide deforestation. He also seems to be devising various methods for depleting the blanket of ozone in the stratosphere. The danger may be slight as yet, but it is not nil, even though the weather and climate are to all intents and purposes stabilized by the oceans which cover two-thirds of the Earth's surface.

Clouds

The yellow colour of the Cytherean clouds has long been cited as evidence that, unlike the white clouds of Earth, they do not consist of water vapour. Detailed measurements obtained by means of the Pioneer probes have now given us an excellent picture of the cloud structures, and also the composition of the various layers.[7,8]

There appears to be a haze layer overlying the main cloud-deck. This haze is simply the outer cloud layer observed by telescopic workers on Earth. It has been identified as being made up of sulphuric acid droplets of radius between 1 and 2 μm, but the haze layer may not be permanent; it has been seen to appear and vanish again over periods of several years.

The middle-cloud region, between 5 and 56 km altitude, shows a decrease in particle concentration to about 100 cm^{-3}, but involves larger liquid droplets and solid particles of radius 10 to 20 μm. The densest layer lies between 49 and 52 km, where visibility is reduced to less than one kilometre. This is the region with the largest particle size. Just below this opaque layer is a distinct thin region of somewhat reduced opacity (Fig. 22).

In the region from 32 to 48 km lies a thin haze of particles almost 1 μm in size, with concentrations of 1 to 20 cm^{-3}. The upper part, from 45 to 47 km, is the most dense. Below the lower boundary, the atmosphere seems to be free of particles all the way down to the surface. However, this does not mean that visibility is unlimited. Venus' atmosphere is so dense that even in the absence of clouds the scattering of light by gas molecules alone probably restricts the limit of vision.

In general, the measurements indicate a concentration of particles either independent of height, or else decreasing at lower altitudes. This behaviour is consistent with a source at the top and a sink at the bottom, with downward flow by mixing and by gravitational settling (Fig. 23). This suggests that there is a source of sulphur and sulphuric acid at the top of the middle-cloud region, at about 57 km. This source is presumably photochemical in nature, requiring ultra-violet radiation which does not penetrate far into the sulphuric cloud. The H_2SO_4 droplets lose water on the way down, but then encounter increased H_2O and perhaps SO_2 and SO_3 as they fall under the influence of gravity, so forming a layer of dilute acid

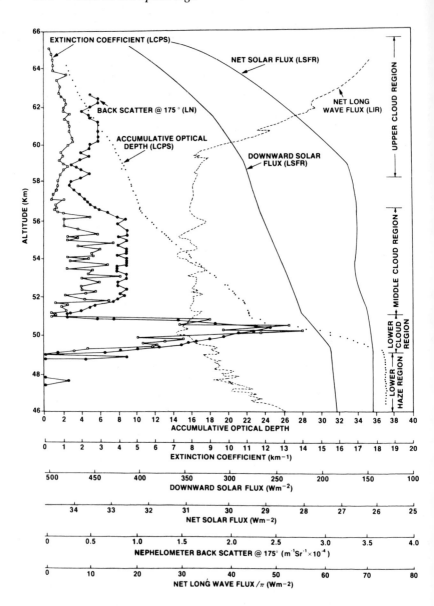

Fig. 22   Structure of the Venus clouds and the measurements of optical thickness, solar and infrared fluxes through the cloud layers.

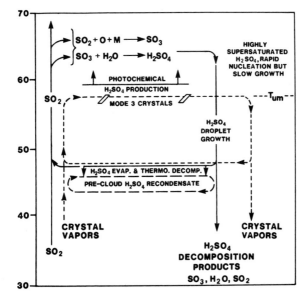

Fig. 23   The 'sulphur' cycle of the Venus clouds.

which is concentrated in the lower-cloud region. Most importantly, the large sizes of the particles (greater than 10 μm) encourage the possibility of precipitation on Venus, either from the dilute H_2SO_4 lower-cloud region or else initiated by sulphur from the middle-cloud region, which would pick up dilute H_2SO_4 when falling below 51 km.

Besides water, there are several substances which are almost non-existent above the clouds but well represented below them. The abundance of SO_2 varies in this way by a factor of more than 200, and oxygen by a factor of 60. These higher abundances arise when sulphuric acid and sulphur rain out of the clouds, vaporize, and are even dissociated in the heat below, so that they chemically attack the gases of the lower atmosphere.

The presence of SO_2 and not COS as the major sulphur-bearing compound in the lower atmosphere is a major surprise. Previously it had been thought that the abundance of COS indicated that the CO abundance in the atmosphere might be affected by certain minerals on the surface, and that its value was therefore representative of the amount existing below the clouds, but the new measurements suggest that CO is generated almost entirely by photochemical processes, and no significant amounts of COS should be expected.

Perhaps one of the most startling findings in connection with the Cytherean atmosphere is the suggestion that lightning occurs at some level beneath the clouds. The Venera 11 and 12 space-craft have detected 13 minutes of electromagnetic signals similar to terrestrial lightning storms; they began at an altitude of 32 km and continued down to only 2 km above the surface. At times as many as 25 lightning-strokes per second were detected.[9] Thirty-two minutes after landing, the Venera 11 acoustic experiment detected a noise level of 82 decibels, presumably due to thunder. In addition, the Pioneer orbiter detected 100 Hz bursts at low altitude, which were atmospheric in origin and are consistent with the lightning theory.[10]

Should we expect thunder and lightning in the atmosphere of Venus? We naturally know a great deal about the causes of such phenomena in our own atmosphere. In general, we require the formation of crystals, which results from the production of positive charges in the water-droplets as they freeze, but rapid vertical motions in the atmosphere are also needed. We have no direct evidence for large particles or strong updraughts on Venus. Therefore, if lightning really does occur in the Cytherean atmosphere, we must assume that large particles must exist even though we have not yet been able to detect them. Of course, lightning may be associated with volcanic activity. Pioneer Venus results indicate intense lightning over the volcanoes in Beta Regio which may well be erupting now.

Another problem concerns the origin of the ultra-violet features that have been regularly observed in the clouds. Two possible explanations have been suggested from the Pioneer results. The first is the assumed presence of gaseous SO_2, which is highly opaque at wavelengths less than 3200 Ångströms.[11] The second source dominates the longer wavelengths. One likely candidate is gaseous CI_2; approximately 1 ppm of it would be needed to account for the observed spectral contrasts.[12] Whether this amount of CI_2 could be generated from the photodissociation of HCl is not clear, since it depends upon the precise amount of HCl and the other minor constituents of the gases in the region of the upper clouds.

Cloud Morphology and Motions

Although Venus usually appears as a featureless disk, observations at ultra-violet wavelengths are extremely rewarding, and show a wide range of cloud structures. Indeed, it was these observations which first showed up the very rapid rotation of the upper atmosphere of Venus. Now, from the Pioneer Orbiter, the cloud systems have been studied continuously for a period of more than two years,[13] and we have been able to gain a great deal of information about these features, which are so important in Venus' unique meteorology (Plate 10). The planetary scale and the small-scale markings shown in the images provide information about the horizontal and vertical cloud structure, atmospheric waves, and wind velocities at the cloud-top level (Fig. 24).

Infra-red observations from the Pioneer Orbiter have led to the discovery of a significant cloud morphology in the north polar region of Venus, which appears as a dipole structure.[4] It consists of two clearings in the clouds, at locations straddling the pole and rotating around it in about 2·7 days. The clearings are thought to be evidence for subsidence of the atmosphere at the centre of the polar vortex. The absence of descending motion elsewhere suggests that a single large circulation cell may fill the northern hemisphere at levels near the cloud tops (Figs. 24 and 25). A crescent-shaped collar region consisting of anomalous and variable temperature and cloud structures surrounds the pole at about 70°N, and rises to perhaps 15 km above the mean height of the cloud tops. It has a fixed solar component, and sometimes shows spiral breaks. This feature and the double vortex eye are large, persistent deviations from the mean circulation, due to planetary-scale waves of unknown origin.[4]

The polar regions were found to be appreciably brighter during the first phase of the Pioneer Venus mission than they had been during the brief Mariner 10 fly-by. The planetary-scale dark Y-feature, previously identified both by Earth-based telescopic observers and on the Mariner 10 images,[14] rotates around the planet with a period of from 4 to 5 days, but often changes dramatically from one presentation to the next. Sometimes it even disappears completely.[13] Bow-shaped features and also cellular features occur at all longitudes, but are more easily observed near and downwind from the subsolar point. The pattern seems to be composed of both diffuse bright cells 500 to 1000 kilometres in diameter with dark rims, and dark cells with bright rims. All the cloud features show a globally co-ordinated oscillation of their orientation relative to latitude circles. The early Pioneer observations again show zonal winds of the order of 100 m/second near the equator, as also had

10

A sequence (left to right, top to bottom) of nine ultra-violet images of Venus commencing on 10 February 1979. Each image, obtained by the Pioneer Venus Orbiter Imaging System, is for an LT of about 0900 so the sequence covers an 8-day period. The 4-day rotation of the upper clouds is clearly seen.

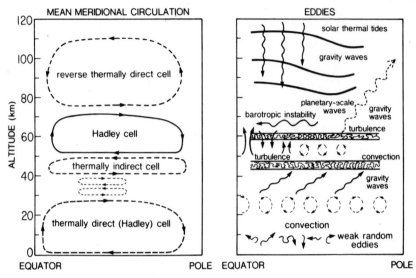

Fig. 24   Schematic of the Venus mean meridional circulation and atmospheric eddy phenomenon.

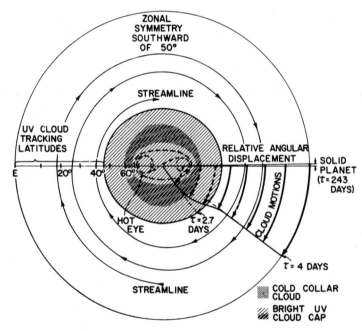

Fig. 25   The flow around the polar vortex observed in a polar stereographic projection.

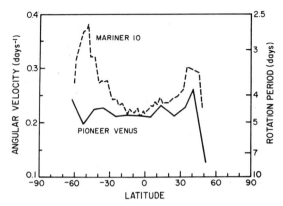

Fig. 26 Variation of the rotational period as a feature of latitude at the top of the visible clouds from Mariner 10 and Pioneer Venus observations.

been indicated from Earth-based and Mariner 10 studies. However, important differences are found from the earlier investigations (Fig. 24). Mariner 10 showed two strong mid-latitude jets at 35°S and 15°N, and a stronger jet at latitude 45°N. These results demonstrate the natural variability of the meteorology of Venus.

Circulation of the Atmosphere

Without doubt, the most surprising discovery of all has been the rapid rotation of the atmosphere of Venus at the level of the cloud tops. The rotation period here is only 4 days, which is very fast compared with the solid-body rotation of 243 days. Indeed, Venus possesses the largest ratio between atmospheric and planetary rotation rates of any body in the Solar System.

What mechanism or mechanisms can be capable of generating these tremendous winds, which blow in the opposite direction to, and twenty times faster than, the overhead motion of the Sun relative to a fixed point on the surface of Venus? What are the motions at other levels in the atmosphere, both above and beneath the clouds?

Almost all the solar energy is absorbed in the neighbourhood of the cloud tops, so providing the heating which drives the observed motions.[15] Its effects are conveyed to higher levels by transport processes. In the stable upper atmosphere of Venus, internal gravity waves play a major rôle in the vertical transport of heat (Fig. 24). Since these mechanisms do not act simultaneously, the horizontal movement of the source of heat produces tilt in the vertical pattern of convection which results in a net motion in the direction opposite to that of the Sun. It is possible to create mean motions much faster than the speed of the heat source. The magnification factor is largely determined by the deviation of the vertical lapse-rate of the temperature from the value drawn from the distribution of heating and cooling in the Cytherean atmosphere.[16]

The high-resolution images of Venus show many examples of small-scale structures in the clouds. These whirls and swirls are generally referred to as eddies, which, typically, cover a wide range of spatial scales (Fig. 24).[13] They include convection, small-scale gravity waves and planetary-scale waves, covering a range from several kilometres up to that of the planet itself. Eddies are very important for the transfer of energy and momentum in planetary atmospheres, and this is certainly the case with Venus.

The dominance of the largest-scale eddies is evident in those images which are so often characterized by the single dark Y-feature which encircles the planet. This feature probably results from the superposition of two large-scale waves whose individual parts drift slowly in and out of phase before dissipating. This situation may well explain the change in shape from a Y to a C, and the occasional absence of the feature.[13,17,18]

In other regions the picture is more complicated. Velocities increase to either side of the equator, and at a latitude of approximately 50° the rotational period may be as short as 2 days. Streaks in the clouds show a flow towards the polar vortices, where the kinetic energy of rotation is eventually dissipated.

To maintain the planetary angular momentum balance, the return flow of the atmosphere occurs at deeper levels in the atmosphere. Fig. 9 shows a graph for these motions which is consistent with the temperature and velocity measurements of the individual probes. The Hadley cells are essential characteristics of the flow, providing the latitudinal redistribution of heat in the atmosphere. The winds decrease rapidly in the atmosphere nearer the surface (Fig. 27)—from a value of 100 m/second at the cloud-top level. At 50 km the velocities are about 50 m/second, decreasing to only a few metres per second near the surface. The overall situation seems to reduce to a zonal retrograde rotation dominating the circulation between one and two scale heights above the clouds. Superimposed on this atmospheric rotation is a cloud-level pattern of Hadley cells. Although the circulation shows unexpected

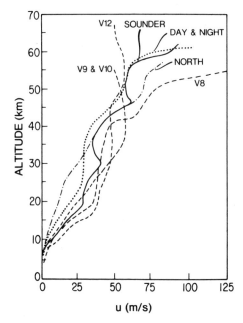

Fig. 27
A comparison of the general
wind velocity profiles from
Pioneer Venus and Venera
probes.

variability with time, the large zonal wind motions as well as the small poleward meridional motions at the cloud tops are persistent features. It would seem that the Hadley cells or eddies are essential for the redistribution of heat in the deep atmosphere, although it is difficult to say just how many of these features are present (Fig. 24).

It should also be noted that the stratosphere of Venus is not clearly defined. On the Earth, this region is created through the absorption of solar energy by ozone, but there does not seem to be an equivalent process taking place on Venus.

The upper-atmosphere motions are particularly interesting; we know much more about them now than we did before the flight of Pioneer Venus. The zonal winds decrease from 100 m/second at the cloud tops to less than 50 m/second at an altitude of 100 km.[3] In the stratosphere and mesosphere regions, vertical and horizontal propagation of planetary-scale waves transports heat polewards in a manner similar to that of the Earth's atmosphere. A significant difference between the two atmospheres is that for Venus, the thermal structure of the stratosphere is affected by the morphology and the radiative properties of the ubiquitous clouds.[4] Rather surprisingly, the upper atmosphere of Venus seems to be rather insensitive to solar activity, suggesting that this region is not controlled by the usual heating mechanisms associated with planetary thermospheres. We still have a great deal to learn.

But why is the meteorology of the upper atmosphere of Venus so different from that of the Earth? The rapid rotation of the upper atmosphere requires an efficient high-altitude thermal mechanism, but for the Earth the greatest amount of available radiative energy is that absorbed at the surface. Could other planetary atmospheres show this excess rotation? Certainly any atmosphere where heating tends to produce an equatorial thermal bulge, but which cannot develop instabilities because of slow rotation or powerful damping, must develop an excess rotation at a high level, in the same sense as the rotation of the solid globe. This may well be the case with Titan, the largest satellite of Saturn, which we now know to have a dense, massive atmosphere—though the main constituent is not carbon dioxide, as with Venus, but nitrogen.

Conclusion

Without doubt, the present-day discoveries with regard to the atmosphere of Venus need very careful examination. The basic composition of the atmosphere has provided new insight into the formation of the Solar System. Measurements of the deep atmosphere have endorsed the theory of the runaway greenhouse effect, and have also underlined the fundamental problems of atmospheric pollution. The atmosphere has proved to be strikingly different in structure from that of the Earth. It is hot at the bottom and cold at the top, in complete contrast to that of the Earth, which is cold near the surface and at a high temperature at great altitudes. The surface temperature of Venus is about 450°K higher than that of the Earth, whereas the Earth's upper atmosphere is about 800°K 'hotter' than that of Venus.

With regard to weather systems, it has been found that Venus' circulation is not a simple system related to a subsolar circular flow, as was previously thought. The quick rotation of the uppermost atmosphere makes up a significant part of the circulation, coupled with the energy transported with the eddies and the Hadley-type cells. We have learned much from these multiprobe missions, but many important questions remain to be answered.

12 The Surface of Venus

Each planet in the Solar System has its own peculiarities, and each presents the observer with its own special problems. Of the inner planets, Mercury[1] and Mars have atmospheres which are either negligible (in the case of Mercury) or are generally more or less transparent even though appreciable (as with Mars), so that maps can be drawn up. Mars is much the easier target, partly because it is closer to us than Mercury, partly because it is larger, and partly because it is further away from the Sun than we are, so that when best placed (near opposition) the whole of the Earth-turned hemisphere is in sunlight. The first reasonably accurate maps of the Martian surface were produced over a century ago, and it was possible to measure the rotation period to within an accuracy of a fraction of a second even without the help of radar or space-probes. Now that we have very detailed charts of the entire surface, drawn from the Mariner and Viking results, it has been confirmed that the visual maps drawn up in pre-Space Age were very satisfactory.

Mercury[1] is a difficult object to study; surface details are not easy to make out. Probably the most concentrated programme of mapping before the flight of Mariner 10 was undertaken by E. M. Antoniadi, in the 1920s and 1930s, with the aid of the Meudon 83-cm refractor. Unfortunately, the problems proved to be insurmountable, and Antoniadi's chart[1] bears little resemblance to the actual surface topography as revealed by Mariner 10, while later attempts by Dollfus and others were no better. This is no reflection upon the observers themselves, and it must be admitted that the completion of our maps of Mercury must await the launching of further probes.

Venus is different again. As we have seen in Part 1, attempts to draw up permanent maps were unsuccessful—and had to be so, because only the upper clouds can be seen from Earth, and the features are always changing. The cusp-caps were known to be of interest, and the Y-feature seen by French observers and others was persistent, but that was all. It was only with the advent of radar that it became possible to start mapping the Cytherean surface.

Hot and dry, and under a tremendous atmospheric pressure more than 90 times that of the Earth, the surface of Venus is unique in the Solar System. For many years the lower levels remained quite unknown; they were veiled by the dense sulphuric-acid cloud layers. But now, through advances in radar technology, it is possible to penetrate these massive clouds, and to investigate the surface structures of Venus.[2]

The radar echoes from the planet are dominated by specular reflection, rather as sunlight glints from the wavelets on the ruffled surface of a sea or lake. The rate of decrease in the power of the signal returned to the detector on Earth, or on a space-craft, varies, so that the viewing geometry of the observations may be related directly to the steepness of the slopes of the feature observed. However, the total power of the scattered signal depends on the reflectivity of the surface. Irregularities on the surface, or features with sharp edges, can easily be detected by the diffuse scattering of the radar beam spread over a wave range of emerging angles. Now, these techniques, which have been applied to Venus at the Arecibo Observatory and at the Goldstone tracking facility, have been greatly extended by regular observations from the orbiting Pioneer Venus space-craft.[3] This has provided almost global coverage of the surface, and has revealed a surprising landscape—previously hidden beneath the clouds.

Venus itself is quite spherical. This is very different from the other planets, and also from the Moon. Venus does not appear to bulge at its equator, and neither is it flattened at the poles. The surface of the planet is in general quite flat. There are a few large continents and some smaller island-type areas, which rise above the global plain of the planet (Plate 12). This topographical distribution is reflected in the narrow range of surface elevations on Venus. About 20 per cent of the planet lies within 125 metres of the average radius of 6051·4 km; 4 per cent lies within 500 metres of this level, while a further 90 per cent lies within a range of 3·5 km. This is in contrast with the surface of the Earth, which varies considerably in level between continents and oceans. On Earth, about 35 per cent of the surface lies within the outer boundary of the continental shelves, while the remaining 65 per cent covers the ocean floor area. On Mars we have a picture which is again totally different; the huge highland areas of the Tharsis ridge occupy a significant part of the surface.

The Rolling Plains

The distribution of topography should not be taken to suggest that Venus is completely flat (Fig. 28). Actually, the planet does have some massive mountains. The highest point on Venus is in the highland area, and has been named Maxwell Montes (latitude 63°.8N, longitude 2°20 (Figs. 29, 30, Plate 12). It is 11·1 km above the average level of the planet, so that it is considerably higher than Mount Everest on the Earth, which attains only a modest 8·8 km. The lowest point on Venus is the rift valley named Diana Chasma (latitude 14°S, longitude 156°), which is 2 kilometres below the mean radius of the planet. Compared with terrestrial depths, Diana Chasma is deeper than the Dead Sea Rift, but is less than one-fifth the maximum depth of the Mariana Trench, which is 11 kilometres below sea-level. The depth of the Diana Chasma below the adjacent ridges is 4 kilometres, which is approximately half the depth of Valles Marineris, the huge canyon system on Mars (Fig. 30).

Venus has three main geological regions: the upland plains, the lava-filled plains, and the continents. Approximately 70 per cent of the planet's surface consists of rolling upland plains, which rise 2 kilometres above the mean level, and on which there are circular dark features. The altitude of these plains is similar to that of the Martian upland areas, and also to the lands on Earth which are elevated only about 800 metres above sea-level.

Circular features with diameters between 20 and 300 kilometres were first recognized in the early Goldstone images of Venus.[6] Subsequent observations from the Pioneer Orbiter have shown a range of circular features of about 500 to 800 kilometres in diameter (see Fig. 31).[3] However, all of these seem to be very shallow, with depths of only 200 to 700 metres. Some of the surface features may be of either volcanic or impact origin; certainly a great deal of cratering took place on all bodies during the early history of the Solar System. An interesting feature is Sappho, centred at latitude 70°N, longitude 15°, which is about 200 to 300 kilometres in diameter; it is situated on a region slightly higher than its surroundings, and is elevated, with linear features radiating from it, which suggests a probably volcanic origin. Perhaps the dark centres represent summit calderas; the linear features may be similar to the volcanic flows found on other parts of the planet—for instance, in the region of Beta Regio.

Some of the bright rings which surround many small dark areas in

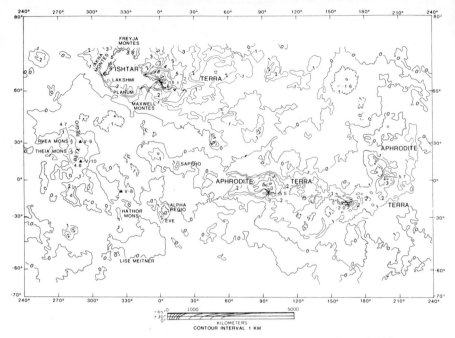

Fig. 28 Generalized topographic map of Venus, with a contour interval of 1 km. The positions of Venera 8.9.10 spacecraft are shown as V8,9,10.

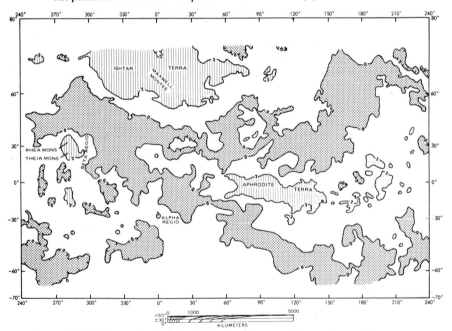

Fig. 29 A map showing the distribution of topographic provinces on Venus. Rolling plains (clear) 0–2 km; highlands (hatching) 72 km, and lowlands (dotted pattern) < 0 km, relative to a radius of 6051 km.

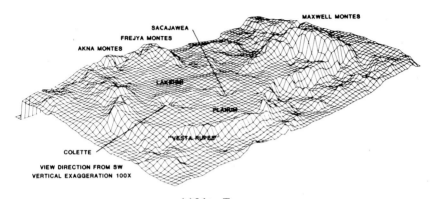

(a) Ishtar Terra

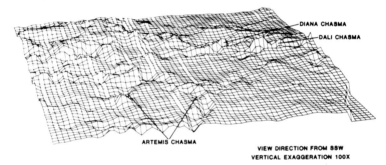

(b) Rift valleys within Aphrodite Terra

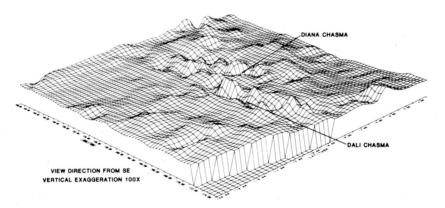

(c) Rift valleys of the Aphrodite region

Fig. 30   Diagrams to illustrate the terrains on Venus

the rolling plains are wider than most ejecta blankets around normal impact craters. These features may well be volcanic in origin. On the Earth, Moon and Mars, many authorities consider that impact craters can be distinguished from volcanic craters by the characteristics of their surrounding ejecta deposits, though there is still considerable disagreement as to the main cratering process which operated on the Moon and Mars. For Venus, the Earth-based and Pioneer radar images are too low in resolution for such features to be detected.

It is not possible to decide whether wind-blown material occurs in the Cytherean craters. However, the Venera 9 and 10 pictures (see Plate 11) do show fine-grained dark material which is possibly wind-blown.[7] The windspeeds near the surface are of the order of 1 to 2 m/second, which could be sufficient to move some sedimentary material.[8]

11 First views of the surface of Venus. The top picture obtained from Venera 9, and the lower one from Venera 10. The horizon can be seen in the top right-hand corner of each picture. Detail includes angular boulders with smaller particles nearby.

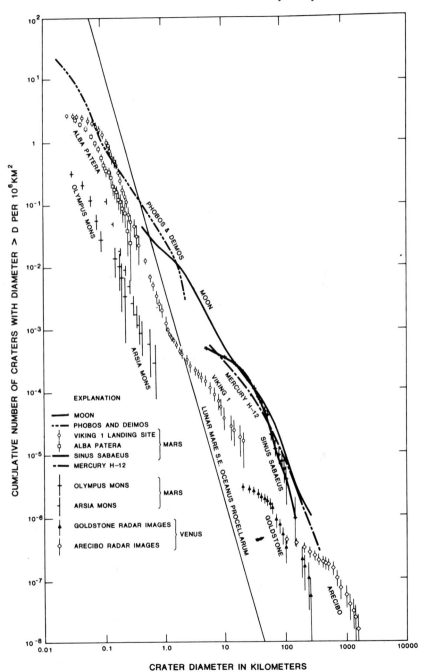

Fig. 31 Crater density counts compared with counts derived from
Goldstone and Arecibo images of the Venus surface with data
from Moon, Mars, Mercury, Phobos and Deimos.

A comparison of the crater densities of Venus with those of the other terrestrial planets (Mercury, Earth and Mars), together with the ancient regions of the Moon and the Martian satellites, indicates considerable similarity (Fig. 31). This supports the view that Venus has a heavily cratered ancient crust. Venera 8 landed within the rolling plains province, and measured a granitic composition composed of uranium, thorium and potassium content in the surface soil.[9] This also is in agreement with the general geological interpretation. Rocks of the rolling plains may be anorthositic in composition, as with some of the lunar soils.

The Lowlands

The lowlands of Venus represent about 10 to 20 per cent of the surface; they differ markedly from those which cover about 70 per cent of the surface of the Earth.[3] An extensive lowland area shaped like a letter X lies south of Ishtar Terra, between Aphrodite Terra and Beta Regio (Fig. 29). The region known as Guinevere Planitia is in the north-west of the X, and is the most rectilinear of the lowland areas. The other lowland regions of the planet are quite isolated, and form discrete patches among the other regions of the Cytherean surface.

East of Ishtar Terra the surface descends to the most extensive basin on the planet, Atalanta Planitia, centred at latitude 65°N, longitude 165°. Atalanta is about the size of the Gulf of Mexico, and lies 1·4 km below the mean radius. The lowest point of this basin is actually at 1·6 km below the mean radius, and is therefore nearly as low as the deepest point of Diana Chasma.[4] Atalanta appears to be very smooth, and resembles the lunar mare basins, the Martian northern plains, and the ocean basins of Earth. No large craters have yet been detected in Atalanta, or in other lowland regions, so that these smooth, uncratered areas may be basaltic lava flows similar to those which fill the mare basins on the Moon and the ocean floors of the Earth. Certainly the basaltic composition of the rocks measured near Beta Regio seems to support this conclusion.

The lowland surface is probably young. Gravitational anomalies indicate that the plains have a thin crust, of low density compared with that of the crust under the upland plains—a situation found also with the Moon and Mars.[10]

The Highlands

Although the highland regions make up less than 8 per cent of the surface of the Venus that has so far been observed, they seem to dominate the topographic maps of the planet.[4] The highlands occur in three main areas: Ishtar Terra (latitude 65°N, longitude 350°), Aphrodite Terra (on and south of the equator between longitudes 60° and 205°), and Beta Regio (latitude 15° to 40°N, longitude 275° to 296°) (Plate 12). Two of these regions are as large as terrestrial continents. Aphrodite is the size of Africa, while Ishtar is equal in area to Australia. Beta Regio appears smaller, and may differ from the larger continents in both age and composition.

Ishtar Terra is about 11 kilometres, and Aphrodite Terra about 5 kilometres, above the mean level of the Cytherean surface. They may represent lava-covered uplifted segments of low-density crust, surmounted by large volcanic structures composed of intermediate or silic rocks. Volcanic and tectonic features seem to be more clearly defined on Ishtar than on Aphrodite.[5] These characteristics are shown on the images as well-developed ridge and trough systems, and lineaments and scarps associated with elevated terrains. The degraded appearance of Aphrodite may indicate that it is older.

The Ishtar Terra region also contains the largest peaks of the Cytherean system. They are made up of three units, the Maxwell Mountains, Lakshmi Planum, with the mountain ranges of Akna Montes and Freyjas Montes on its northern and north-western regions, and an extension of the Lakshmi Planum. The Lakshmi region is from 4 to 5 kilometres above the mean radius of the planet; this is much the same level as that of the Tibetan plateau above sea-level on Earth. It is, however, about twice the size of its terrestrial counterpart. On its southern boundary there is a bright scarp which may consist of slopes of eroded débris caused by slumping and faulting. If Ishtar consists of basaltic lava-flows, it should create a gravitational anomaly, but no such anomaly has been detected, suggesting that the Lakshmi plateau more probably consists of thin lavas overlying an uplifted segment of ancient crust similar to the structure of the Tharsis region of Mars.[10]

Maxwell Montes itself is situated on the eastern side of Ishtar (Plate 12). At the top of this great feature there is a caldera 100 kilometres across and 1 kilometre deep, offset to the eastern flank of the volcano, some 2 kilometres below the actual summit. The slopes of Maxwell are steep, and seem to be covered with very coarse

12   A schematic of Ishtar Terra, a large continental-sized plateau compared with the continental United States. The highest point yet found on Venus is Maxwell Montes. On the east flank of Maxwell is a dark circular feature more than 900 m deep which may be a volcanic caldera.

débris. This interpretation is strengthened by the Venera 9 and 10 pictures, showing rocky surfaces at their landing sites.

The other main continental region, Aphrodite Terra, lies near the Cytherean equator. It is about the size of Africa and contains two mountainous areas. To the east, the mountains are 4 kilometres above the mean radius of the planet, while the mountains to the west are twice as high. Between these ranges is a vast plain.

Since these continental areas do not yet show any circular features that could be craters, they are assumed to be geologically very young. The existence of the continents also implies that there can be little water in the crust of Venus. At the high temperatures of 737°K, any water-rich crustal rocks would easily deform, and the continents would not survive over geological time-scales.

The small highland area of Beta Regio is dominated by the shield volcanoes Theia Mons and Rhea Mons, together with a large trough, along either side of which are smaller volcanoes. Theia and

Rhea are 4 kilometres high, and are probably active at the present time. The trough is part of a fault zone which extends further southward, where two other small highland areas are aligned. Lava-flows spread out to the east and west; the Venera 9 and 10 space-vehicles landed in these flows, confirming their basaltic nature. Bright streaks radiate from the shield volcanoes; these suggest lava-flows, which in turn suggests that the region is comparatively young.

kilometres above the mean level, and lies in an area roughened by many fractures. This general region resembles the Tharsis area of Mars; it may well be an old volcanic region.

East of Ishtar there is a large region, extending from 50° to 75° latitude and 4° to 14° longitude, which consists of complex ridges and troughs, probably disrupted by extensive faulting. It seems to be the most tectonically disturbed region of Venus (Plate 13). The region to the east and south-east of Aphrodite is also particularly chaotic. A roughly circular feature 1800 kilometres across, centred at latitude 35°S and 135° longitude, may be the faint remnant of a gigantic crater.

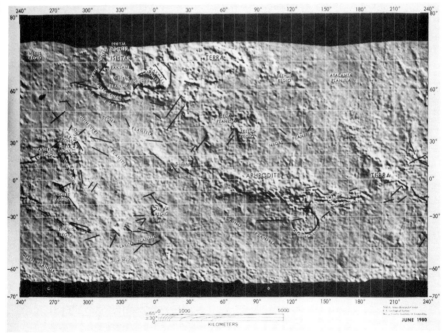

13   A map of Venus showing tectonic features which are thought to be scarps, trenches and ridges.

Though Venus is so like the Earth in size and mass, it is a different kind of world. Studies of plate tectonics have become all-important so far as the Earth is concerned, but recent results (mainly from the orbiting section of Pioneer) indicate that Venus, with its thicker crust, is a 'one-plate' planet. This means that its volcanic activity is more localized. The Earth vents its internal heat at many points, but as yet only two areas of active vulcanism have been positively identified on Venus. One is Beta Regio, where the huge volcanoes of Rhea and Theia are thought to stand over a powerful, upflowing convective plume deep in Venus' interior magma. The other is the so-called Scorpion's Tail of Aphrodite Terra. The observed concentration of lightning over these two regions is a reasonably sure sign of common vulcanism there.

The first attempts at analysis of the surface materials were made in March 1982 by the two latest Russian probes, Veneras 13 and 14. Both landed in the general area of Phœbe Regio, adjoining Beta Regio. Each space-craft consisted of a lander and a separate section which acted as a relay for transmissions from the surface of Venus, and then continued in orbit round the Sun. Venera 13 came down first; the lander, braked by parachutes, took just over an hour to reach the ground after separation from the main probe, and continued to transmit for a record 127 minutes after arrival, sending back panoramic pictures (Plate 14) and also the first colour views from the surface. The temperature was given as 457°C., and the pressure as 89 atmospheres. More important still, it carried a drill which obtained samples of the surface material and drew them back into a hermetically sealed chamber for X-ray fluorescent analysis; the chamber temperature was kept at a mere 30°C., with a pressure only 1/2000 of that outside. The general colour of the rock was reddish-brown.

Venera 14 landed four days later, at latitude 13° 15′ S., longitude 310° 9′. The site appeared to be in the nature of a plain, with none of the sharp, angular rocks which had been recorded from its predecessor. The temperature was slightly higher (465°C.) and the pressure slightly greater (94 atmospheres), but the general results were much the same. In the surface rock, highly alkaline potassium basalts were much in evidence—and the sky proved to be bright orange.

The more one learns about Venus, the more interesting it proves to be, but unfortunately American researches have come to a halt. Even the VOIR probe (Venus Orbiting Imaging Radar), scheduled

14   The surface of Venus in the neighbourhood of the Venera 13 spacecraft. The image shows part of the spacecraft and the pail sampling device used to analyse the properties of the nearby terrain.

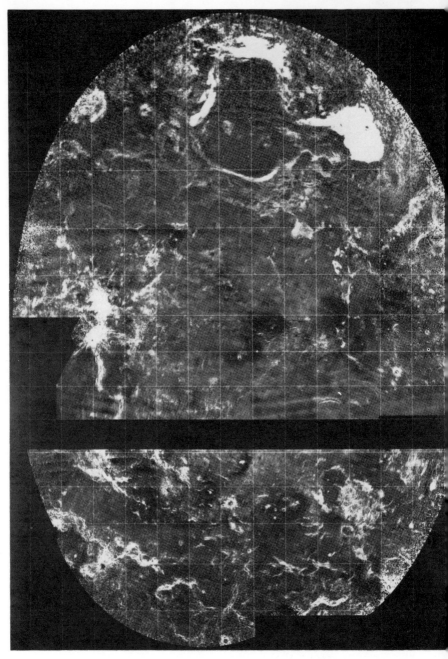

15   Radar images of Venus assembled from observations made at the
Arecibo observatories between 1975 and 1977. The image brightness is proportional
to the degree of surface roughness with dimensions of a few centimetres, except at
low latitudes where **metre scale slopes also influence the reflected** signal.

for the late 1980s, has been a victim of the NASA cutbacks in expenditure, so that our main hopes for finding out more in the immediate future rest with the Russians. They have already announced that their Vega probes to Halley's Comet will fly by Venus and release balloons into the planet's atmosphere, so that the composition and other characteristics will be studied at different levels. It would be wrong to claim that we have solved all the fundamental problems of Venus; there is still much that we do not know, but at least we have made more progress than would have seemed possible only a few years ago.

13 The Interior of Venus

At first sight, the similarity in size, mass and density of Venus and the Earth would suggest that the two planets ought to have similar internal structures. However, from our current understanding of the two worlds, there appears to be some very important difference in their interiors, which in turn has important consequences in their interaction with the planetary environment.

The Earth (see Fig. 32) has a hot, mobile mantle and a crust which is sufficiently thin for it to break up into large plates that move relative to one another. We now recognize these features as the continents. Venus has continents of a sort, and it is natural to ask whether they have any resemblance to their terrestrial counterparts.[1] On Earth, the continents represent regions of gravitational equilibrium with their surroundings. This state, known as isostatic equilibrium, requires that topographical relief be supported by buoyancy arising from a reduced density, rather as an iceberg floats on water. Thus the continents on Earth rise above the ocean basins surrounding them, to an extent which depends primarily on the densities and thicknesses of the continents. The degree of isostatic equilibrium on Venus has been investigated by studying small perturbations in the motion of the Pioneer Orbiter as it passes over large elevated structures. The investigations carried out with regard to the Ishtar and Aphrodite regions have not shown any effects.[2]

Therefore, the large-scale relief on Venus may be in gravitational equilibrium, so that any substantial differences in the density of the crust must be due to the processes of differentiation. On Earth, differentiation has resulted from the large-scale melting of the planet's interior. The dense Cytherean atmosphere suggests that a similar melting has taken place there, through the thorough outgassing of a hot interior. Still further evidence for chemical differentiation on Venus is afforded by the significant amount of radioactivity measured on the surface by the Venera 9 and 10 probes.[3]

There is no evidence of widespread tectonics on Venus. Indeed, the process seems to have been as incipient here as it is on Mars. However, the development on Venus of thin-crusted lowlands and thick-crusted highlands suggests that Venus must have experienced a period of widespread convection in its mantle.

Venus, then, seems to be quite different from the other terrestrial planets. The observed tectonic activity is regional.[1] It is possible that the development of full-scale plate tectonics was stopped by the lack of water and the slow rotation. Taking all these factors into account, the possible interior structure of Venus is shown in Fig. 32. The density of Venus is 2 per cent less than that of the Earth; this is a small but significant difference, which may be due to the processes of formation of the planets.

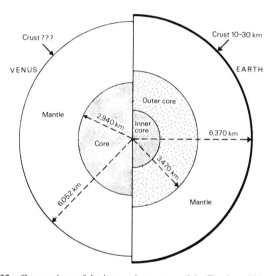

Fig. 32 Comparison of the internal structure of the Earth and Venus.

One of the consequences arising from the fact that the Earth has a large molten iron core is that, because of rotation, the dynamo action generates our magnetic field. However, Venus has no intrinsic magnetic field. On such a slowly-rotating body the dynamo mechanism would create only an insignificant field. The maximum possible strength of the Cytherean field is put at 5×10^{-5} that of the Earth's.[4] Consequently, the solar wind interacts directly with the upper atmosphere and ionosphere of Venus, with complex interactions taking place.

14 Conclusion

In this book, our aim has been to present first an historical survey of Man's efforts to unravel the mysteries of Venus, and then to summarize what has been learned since the beginning of radar and the epic flight of Mariner 2 in 1962. What, then, of the future?

In some ways the future must be regarded as a continuation of the past, and most of the instruments on the Pioneer orbiter are still operating at the time of writing (March 1982). They will continue to monitor the atmosphere, clouds and the surrounding environment for several Cytherean years yet. This is particularly important, and will provide invaluable information on the weather systems of the upper atmosphere of the planet—in particular, the time-history of all the morphological structures in the clouds, which previously have been observed only intermittently. While we can continue to observe the surface of Venus, there is no way in which the spatial resolution of the existing systems can be improved. Observations of the surface at resolutions of 500 metres or even better are urgently required. This would be comparable with the Mariner 9 observations of Mars, which certainly revolutionized our understanding of the planet; earlier, lower-resolution pictures had given the false impression that Mars would turn out to be geologically uninteresting.

There is a great deal still to be learned about the Cytherean atmosphere and meteorology. Future missions are likely to float balloons at different pressure levels, so that by tracking them the winds will be determined. Each balloon package will carry a small set of supporting instruments.

During the next decade we can expect major advances in our understanding of Venus. It is, however, most unlikely that a manned mission will ever be sent there. It would have little value even if it became possible; all necessary investigations can be carried out without the need to protect the human occupants of a space-craft.

16 Venus Orbiting Imaging Radar (VOIR) space-craft which was to have been placed in orbit during the 1980s to investigate the surface structure of the planet.

The conditions which we know to exist on Venus should alert us all, giving us a direct indication of the potential dangers of excessive pollution. We may recoil from the suggestion that the Earth could follow the same evolutionary path as Venus as a consequence of Man's activities, but it is wise to recognize the possibilities. We still have a great deal to understand about our nearest neighbour for the benefit of all men on Earth.

It was once suggested that schemes should be introduced to reduce the carbon-dioxide content of the Venus atmosphere by seeding the atmosphere with suitable chemicals. This approach is unlikely to be successful. The close proximity of the Sun will still provide a large source of energy to re-create the greenhouse effect.

Also, photochemical reactions will continue in the upper atmosphere where the ultra-violet radiation is absorbed. However, there is no harm in speculating: so let us end with an imaginary journey to Venus. What would our space-travellers see, and what would they experience?

The top of the Cytherean ionosphere lies about 400 kilometres above ground level. As we approach, the clouds below appear as a dense haze, totally concealing the great canyons, plateaux and craters which we know to exist. Even from the top of the main atmosphere, at 200 kilometres, the view is much the same. The temperature here is a modest 27°C; at 100 kilometres above ground level it has dropped to −90°C, but then begins to rise again as the atmosphere becomes denser. At 70 kilometres we enter the first clouds, and visibility starts to decrease. By 66 kilometres the Sun is perceptibly dimmed; by the time we have fallen to 63 kilometres the Sun itself is no longer a sharp disk, and is nothing more than a brilliant glare behind the diffuse, yellow cloud-layer made up of tiny particles of sulphuric acid. By now the visibility has fallen to 6 kilometres; the temperature is 13°C, and the outside pressure is about half that of the Earth's air at sea-level.

As we descend, the visibility becomes less, and the temperature rises at an alarming rate. At 50 kilometres the range of view is less than 2 kilometres, and the temperature is 20°C. We pass quickly through a clear layer, and then enter the lower regions; the pressure has risen to 1 atmosphere, while the temperature has soared to 200°C. This is, in fact, the region where the clouds are densest, and only here would they look like the cloud structures on Earth. Beneath this layer, from 47 kilometres or so, we come to a second clear space, with nothing more than what seems to be haze. Even this clears away at 30 kilometres, and the visibility has increased again, reaching 80 kilometres; the illumination is about the same as that on a bright, cloudy day on Earth, but the outside temperature is now a fearsome 310°C. We drop down to 20 kilometres; now the temperature is 380°C, with the visibility down to 20 kilometres once more. The outside light is reddish, and we become acutely conscious of the inferno-like conditions which we are about to meet. From 7 kilometres, some surface features start to become visible in the red gloom below—and finally we land; the temperature now is between 460° and 470°C, and the pressure is a crushing 91 atmospheres. It is not dark, and indeed the overhead illumination is still comparable with that of a cloudy day on Earth, but the visibility is a mere 3 kilometres. The Sun is not visible; it is hidden by the dense atmosphere, and its location can be traced only by an ill-defined, baleful glare. To step outside the space-craft would be to invite disaster. It is obvious to see why Venus has been nicknamed 'the Hell-planet'.

No such journey can be contemplated in the foreseeable future, and indeed may never happen. But in any case, this non-identical twin of the Earth is a fascinating world, and it has much to tell us.

Until very recently, it has been tacitly assumed that life never appeared on Venus, but the new results from Pioneer make us less certain. It is now starting to look as though Venus once had oceans, and perhaps an atmosphere not unlike that of the primitive Earth. Strong evidence for this lost water comes from measurements of the ratio of deuterium ('heavy hydrogen') to ordinary hydrogen. On Venus, the relative abundance of deuterium is as much as 100 times that in the atmosphere of the Earth.

In the early days of the Solar System, the Sun was up to 30 per cent less luminous than it is now. Possibly Venus and the Earth began to evolve along similar lines, with oceans and primitive life. But as the Sun became more powerful, the temperature of Venus rose to intolerable levels; the oceans boiled, and the runaway greenhouse effect began, so that all life on the planet was remorselessly wiped out and Venus became the inferno that it is today. If this is true, then Venus must be regarded as a tragic world. Whether any traces of past life will be found in the rocks, when we manage to secure some and examine them, remains to be seen.

Yet even if life existed on Venus once, it can hardly do so now. Neither does it seem possible that conditions will improve in the future. Therefore, despite all the hopes expressed before the start of the Space Age, it seems that all ideas of reaching Venus and colonizing it must be ruled out—at least in our time.

Appendix 1 Numerical Data

Distance from the Sun:	maximum 109,000,000 km
	mean 108,200,000 km
	minimum 107,400,000 km
Sidereal period:	224·701 days
Orbital inclination:	3°·394
Orbital eccentricity:	0·00678
Sidereal mean daily motion:	1°·60213
Mean synodic period:	583·92 days
Mean orbital velocity:	35·02 km/sec
Diameter:	12,102·8 km
Polar flattening:	Inappreciable
Axial rotation period:	243·01 days
Axial inclination:	178°
Mass, Earth=1:	0·815
Mean surface atmospheric pressure, millibars:	91,000
Reciprocal mass, Sun=1:	408,523·5
Mass in kilogrammes:	$4·8689 \times 10^{24}$
Density, water=1:	5·24
Volume, Earth=1:	0·86
Surface gravity, Earth=1:	0·902
Escape velocity:	10·36 km/sec
Surface Temperature:	737°K (464°C)
Mean visual opposition mag:	−4·4
Albedo:	0·76
Mean apparent diameter of the Sun, as seen from Venus:	44′ 15″
Number of satellites:	0

Appendix 2 Estimated Rotation Periods

There have been many estimates of the rotation period of Venus, and the following list includes most of the more notable attempts. Of course, most of them are of no more than historical interest now, when the problem has been finally solved, but they are worth remembering.

In the third column, V=visual, S=spectroscopic and T= theoretical. The asterisk + indicates that the period was given as captured or synchronous (224d 16h 48m), and an R indicates that the direction was thought to be retrograde. The estimates refer to the solid body of the planet, not to the upper clouds (though until the modern era this was thought, erroneously, to be the same thing).

Year	Observer	Method	Value	Ref
1666–7	G. D. Cassini	V	23h 21m	1
1727	F. Bianchini	V	24d 8h	2
1732	J. J. Cassini	V	23h 15m	3
1740	J. J. Cassini	V	23h 20m	4
1789	J. Schröter	V	23h 21m 19s	5
1801	Fritsch	V	23h 22m	6
1811	J. Schröter	V	23h 21m 7s·977	7
1832	T. Hussey	T	24d 8h	8
1841	Di Vico	V	23h 21m 21s·934	9
1881	W. F. Denning	V	23h 21m	10
1880	G. V. Schiaparelli	V	+	11
1890	Perrotin	V	+	12
1890	Terby	V	V. slow or +	13
1891	Niesten and Stuyvaert	V	24h	14
1891	Löschart	V	23h 21m	15
1892	E. Trouvelot	V	24h	16
1894	C. Flammarion	V	24h	17
1895	Cerulli	V	+	18
1895	Mascari	V	+	19
1895	G. V. Schiaparelli	V	+	20
1895	Tacchini	V	+	21

Year	Observer	Method	Value	Ref
1895	A. S. Williams	V	24h	22
1895	W. Villiger	V	24h	23
1895	L. Brenner	V	23h 57m 36s·2396	24
1896	I.. Brenner	V	23h 57m 36s·27728	25
1897	H. McEwen	V	23h 30m	26
1897	Fontséré	V	+	27
1899	L. Rudaux, G. Fournier	V	24h	28
1900	Müller	V	24h	29
1900	A. Belopolsky	S	24h 42m	30
1901	Vassillieff	V	24h	31
1902	Arendt	V	24h	32
1903	Slipher	S	+	33
1907	Stefanik	V	23h 20–25m	34
1908	Harg	V	Not over 23h 28m	35
1909	P. Lowell	V and S	+	36
1909	Schoy	T	24h	37
1909	Perqueriaux	V	24h	38
1911	A. Belopolsky	S	1d 11h	39
1911	Bellot	T	1d 4h 12m	40
1911	S. Bolton	V	23h 28m 30s	41
1913	Majert	V	24h	42
1915	W. Rabe	V	23h 57m	43
1916	S. Maxwell	T	24h	44
1916	W. F. A. Ellison	T	24h	45
1916	C. E. Housden	V	+	46
1916	D. H. Wilson	V	+	47
1917	H. E. Lau	V	24h	48
1919	W. Evershed	S	20 to 30 h	49
1921	R. Jarry-Desloges	V	22h 53m	50
1921	W. H. Pickering	V	2d 10h	51
1922	H. Kaul	T	1d 2h 18m 53s	52
1922	H. J. Gramatzki	V	24h	53
1922	J. G. Yanes	T	24h 33m	54
1922	A. Rordame	Photo	24h	55
1924	H. McEwen	V	2d 20h	56
1924	W. H. Steavenson	V	8d	57
1924	A. Fock	V	24h	58
1925	Nicolson and St John	S	Very slow	59
1927	F. E. Ross	Photo	30h	60
1928	N. P. Schanin	V	23h 58m	61
1928	N. P. Schanin	V	24h	62
1929	R. Jarry-Desloges	V	24h	63
1929	N. P. Sanjutin	V	+	64
1930	J. Camus	V	+	65
1932	R. Barker	V	+	66, 67
1933	E. P. Martz	V	1d 13h 4m 48s	68

Year	Observer	Method	Value	Ref
1934	E. M. Antoniadi	V	V. long or +	69
1934	L. Andrenko	V	+	70
1936	W. W. Spangenberg	V	24h	71
1939	W. H. Haas	V	+	72
1940	L. Romani	V	+	73
1942	W. H. Haas	V	Slow	74
1945	Phocas	V	Slow	75
1949	Schirdewahn and Schwartz	V	23h 15m	76
1949	V. V. Volkov	V	2d 12h	77
1951	R. M. Baum	V	195d	78
1951	H. Le Vaux	V	1d 7h 12m	79
1952	J. C. Bartlett	V	22h 30m	80
1953	G. D. Roth	V	15h	81
1954	M. Kutscher	V	1d 11h	82
1954	G. P. Kuiper	Photo	A few weeks	83
1955	A. Dollfus	V and Photo	+	84
1956	J. D. Kraus	Radio	22h 17m	85
1958	R. S. Richardson	S	Slow. R?	86
1958	I. I. Gusev	V	22h 30m	87
1962	R. L. Carpenter	Radar	About 250d. R	88
1962	R. M. Goldstein	Radar	248d±50d. R	89
1962	W. K. Klemperer, G. R. Ochs, K. L. Bowles	Radar	180d to 280d	90
1962	D. O. Muhleman	Radar	570d or 250d. R	91
1962	F. D. Drake	Radar	Slow; R.	92
1962	O. N. Rzhiga	Radar	About 300d. R	93
1964	J. E. Ponsonby, J. H. Thomson, K. Imrie	Radar	100 to 300d. R	94
1964	I. I. Shapiro	Radar	274d±4d. R	95
1964	R. L. Carpenter	Radar	258d±6d. R	96
1967	I. I. Shapiro	Radar	243d·09±0d·18	97
1979	I. I. Shapiro	Radar	243d·01±0d·03	98

Note that the 'Earth-locking' rotation period would be 243d·16. If Shapiro's latest value is correct, therefore, the 'lock' is not perfect, and must be dismissed as coincidence. The 'solar day' on Venus is 117 Earth-days; that is to say, a daylight period of 58·5 days followed by a night of equal length.

Appendix 3 Conjunctions and Elongations, 1980–2000

Date	Phenomenon	Diam, sec of arc	Magnitude
1980			
5 Apr	Gr. elongation E 45°9'	24	−4·0
9 May	Gr. brilliancy	37	−4·2
15 June	Inferior conjunction	58	−2·8
22 July	Gr. brilliancy	37	−4·2
24 Aug	Gr. elongation W 45°9'	24	−4·0
1981			
7 Apr	Superior conjunction	10	−3·5
11 Nov	Gr. elongation E 47°2'	25	−4·0
16 Dec	Gr. brilliancy	41	−4·4
1982			
21 Jan	Inferior conjunction	63	−3·1
25 Feb	Gr. brilliancy	40	−4·3
1 Apr	Gr. elongation W 46°5'	25	−4·0
4 Nov	Superior conjunction	10	−3·5
1983			
16 June	Gr. elongation E 45°4'	24	−3·9
19 July	Gr. brilliancy	37	−4·2
25 Aug	Inferior conjunction	58	−3·2
1 Oct	Gr. brilliancy	38	−4·3
4 Nov	Gr. elongation W 46°6'	24	−4·0
1984			
15 June	Superior conjunction	10	−3·5
1985			
22 Jan	Gr. elongation E 47°0'	28	−4·0
26 Feb	Gr. brilliancy	39	−4·3
3 Apr	Inferior conjunction	60	−3·4

Date	Phenomenon	Diam, sec of arc	Magnitude
9 May	Gr. brilliancy	39	−4·2
13 June	Gr. elongation W 45°6′	24	−3·9
1986			
19 Jan	Superior conjunction	10	−3·5
27 Aug	Gr. elongation E 46°1′	25	−4·0
1 Oct	Gr. brilliancy	40	−4·3
5 Nov	Inferior conjunction	62	−3·1
11 Dec	Gr. brilliancy	41	−4·4
1987			
15 Jan	Gr. elongation W 47°0′	25	−4·0
23 Aug	Superior conjunction	10	−3·5
1988			
3 Apr	Gr. elongation E 45°9′	24	−4·0
6 May	Gr. brilliancy	37	−4·2
13 June	Inferior conjunction	58	−2·8
19 July	Gr. brilliancy	37	−4·2
22 Aug	Gr. elongation W 45°8′	24	−4·0
1989			
5 Apr	Superior conjunction	10	−3·5
8 Nov	Gr. elongation E 47°2′	25	−4·0
14 Dec	Gr. brilliancy	21	−4·4
1990			
19 Jan	Inferior conjunction	63	−3·1
22 Feb	Gr. brilliancy	40	−4·3
30 Mar	Gr. elongation W 46°5′	25	−4·0
1 Nov	Superior conjunction	10	−3·5
1991			
13 June	Gr. elongation E 45°3′	24	−3·9
16 July	Gr. brilliancy	37	−4·2
22 Aug	Inferior conjunction	58	−3·2
28 Sept	Gr. brilliancy	38	−4·3
2 Nov	Gr. elongation W 46°6′	25	−4·0
1992			
13 June	Superior conjunction	10	−3·5

Date	Phenomenon	Diam, sec of arc	Magnitude
1993			
19 Jan	Gr. elongation E 47°1′	23	−4·0
24 Feb	Gr. brilliancy	39	−4·3
1 Apr	Inferior conjunction	60	−3·4
7 May	Gr. brilliancy	39	−4·2
10 June	Gr. elongation W 45°8′	24	−3·9
1994			
17 Jan	Superior conjunction	10	−3·5
25 Aug	Gr. elongation E. 46°0′	23	−4·0
28 Sept	Gr. brilliancy	40	−4·3
2 Nov	Inferior conjunction	62	−3·1
9 Dec	Gr. brilliancy	41	−4·4
1995			
13 Jan	Gr. elongation W 47°0′	25	−4·1
20 Aug	Superior conjunction	10	−3·5
1996			
1 Apr	Gr. elongation E 45°9′	24	−4·0
4 May	Gr. brilliancy	37	−4·2
10 June	Inferior conjunction	58	−2·7
17 July	Gr. brilliancy	37	−4·2
19 Aug	Gr. elongation W 45°8′	24	−4·0
1997			
2 Apr	Superior conjunction	10	−3·5
6 Nov	Gr. elongation E 47°2′	25	−4·0
12 Dec	Gr. brilliancy	41	−4·4
1998			
16 Jan	Inferior conjunction	63	−3·1
20 Feb	Gr. brilliancy	40	−4·3
27 Mar	Gr. elongation W 46°5′	25	−4·0
30 Oct	Superior conjunction	10	−3·5

Date	Phenomenon	Diam, sec of arc	Magnitude
1999			
11 June	Gr. elongation E 45°3′	24	−3·9
14 July	Gr. brilliancy	37	−4·2
20 Aug	Inferior conjunction	58	−3·2
26 Sept	Gr. brilliancy	38	−4·3
30 Oct	Gr. elongation W 46°5′	25	−4·0
2000			
11 June	Superior conjunction	10	−3·5

Appendix 4 Venus through the Telescope: Methods of Observation

Venus is a difficult object to study telescopically. When it shines brilliantly down from a darkened sky, low over the horizon, there is little point in observing it at all except for detection of the Ashen Light. Good views may often be obtained during dawn or twilight, but generally speaking it is best to observe Venus in broad daylight, when the Sun is above the horizon. With a clock-driven, equatorially-mounted telescope this presents no difficulty. High magnification is by no means always an advantage. With Venus, the essential is to obtain a sharp, clear image. Very small telescopes are of little use, and an aperture of 15 cm or so is probably the minimum usable for valuable results. Despite the wide range in apparent diameter, it is usually best to make all drawings to the same diameter. A scale of 5 cm to the full diameter of the planet is suitable.

A filter is often helpful. A Kodak Wratten 15 yellow filter is widely used; this has a passband of about 5100 Ångströms, and downward into the red.

Points to be noted include: phase (attempts to determine the value of the Schröter effect, particularly near dichotomy); dark shadings; bright areas; cusp-caps; any irregularities in the terminator, though great care must be taken to differentiate between 'ripples' due to the Earth's atmosphere, and genuine deformations in the Cytherean terminator; any exceptional phenomena, such as bright points; and any sign of the Ashen Light. For the Ashen Light studies, it is essential to block out the bright crescent by means of an occulting bar or some equivalent device, and members of the British Astronomical Association also use a Wratten 35 purple filter with passbands at 4100 and 6600 Ångströms downward.

Venus is also a difficult object for photographic work, but some valuable experiments, using colour filters, have been carried out by Commander H. R. Hatfield; see his papers in *J.B.A.A.*, Vol. 79, p. 229; Vol. 80, p. 59; Vol. 85, p. 42. See also various notes in the 'Report of Observation of the Planet Venus, 1956–72', *B.A.A. Memoirs*, Vol. 41, (December 1974).

References

Chapter 1: **The 'Evening Star'**

1. HOUZEAU and LANCASTER: *General Bibliography of Astronomy*, Vol. I, p. 84.
2. MASPERO: *Histoire Ancienne des Peuples de l'Orient Classique*, p. 95 (1899). See also E. M. ANTONIADI, *L'Astronomie Égyptienne*, p. 94 (Paris 1934).
3. M. JASTROW: *Religious Beliefs in Babylonia and Assyria*, p. 221 (1911).
4. See HINCKS: 'On Certain Babylonian Observations of Venus', *Month. Not. R.A.S.*, Vol. 20, 319 (1860).
5. See J. K. FOTHERINGHAM and H. S. LANGDON: *The Venus Tablet* (1928); also SYDNEY SMITH, *Alalakh*, a pamphlet of the British Museum (1940) and a paper by A. G. SHORTT, *J.B.A.A.*, Vol. 57, 208 (1947). A very valuable survey has been given by J. D. WEIR, 'The Venus Tablets: a Fresh Approach', in *Journal for the History of Astronomy*, Vol. 13, part 1, pp. 23–49, which is the most detailed account so far produced.
6. WILLIAMSON: *Religious and Cosmic Beliefs of Central Polynesia*, Vol. 2, p. 242.
7. R. LINTON: 'The Thunder Ceremony of the Pawnee and the Sacrifice to the Morning Star', compiled from notes by G. DORSEY. Field Museum of Natural History, Department of Anthropology, Chicago (1922).
8. E. M. ANTONIADI, *L'Astronomie Égyptienne*, p. 97 (Paris 1934).
9. HOMER: *Iliad*, XXII, 318. See also XXIII, 226.
10. Ptolemy's *Almagest* was translated into French by M. HALMA (Paris 1813). The movements of Venus are described in Vol. 2, pp. 193–209.
11. C. H. SMILEY: *Jnl. Roy. Astr. Soc. Canada*, Vol. 54, 222 (1960) and *Nature*, Vol. 188, 215 (1960). From his studies of the Venus calendar, Smiley concludes that the Maya possessed consider-

able astronomical knowledge by the fourth century AD—considerably earlier than has been generally supposed.

12. Copernicus' book was published in limited facsimile edition by Macmillan, London 1972. An English translation by E. ROSEN, *On the Revolutions*, was published by Macmillan in 1978.

13. GALILEO: *Sidereus Nuncius* (The Sidereal Messenger), 1610, translated in 1880 by E. S. Carlos. Extracts from it are given in H. SHAPLEY and H. HOWARTH, *Source Book of Astronomy* (New York 1929).

14. *Opere di Galileo*: Vol. 2, p. 42 (Padova 1744).

15. GALILEO: *Dialogue Concerning the Two Chief World Systems* (1632). See pp. 321 and 334 of the translation by STILLMAN DRAKE, University of California Press (1953).

16. STILLMAN DRAKE, op. cit. p. 334.

For further references to Venus in ancient legend, see works such as C. FLAMMARION, *Les Terres du Ciel*, pp. 216–30 (Paris 1884); G. C. LEWIS, *Astronomy of the Ancients*, p. 62 (London 1862). For some fascinating Polynesian legends, see M. W. MAKEMSON, *The Morning Star Rises* (Yale 1941). On p. 140, for instance, Makemson quotes the old Maori belief that the proximity of Venus to the crescent Moon affects the failure or success of a siege.

Chapter 2: **Venus as a World**

1. C. W. TOMBAUGH and PATRICK MOORE: *Out of the Darkness: the Planet Pluto* (London 1980).

2. For further notes on the albedo of Venus, see I. A. PARSHIN, *Astr. Circ. USSR*, No. 145, 12 (1954) and *Bull. Leningrad Univ.* 9, No. 5, 85 (1954). Parshin's values are 0·67 photographic, 0·93 red, 0·95 infra-red.

3. F. ARAGO, *Astronomie Populaire*, Vol. II, p. 533. Quoted in *Popular Astronomy*, Vol. 1, p. 701.

4. N. LOCKYER: *Nature*, 22 December 1887.

5. T. and W. LOCKYER: *Sir Norman Lockyer*, p. 130 (London 1928).

6. P. M. RYVES: *J.B.A.A.*, Vol. 58 p. 127.

7. A. J. L. MURRAY: personal correspondence with Moore, 1958.

8. PLINY: *Naturalis Historia*, II, 6, 8.

9. SIMPLICIUS: *Commentary upon the Heavens of Aristotle*, II, 12,

226a. See also E. M. ANTONIADI, *L'Astronomie*, Vol. 41, p. 343 (1927).

10. E. M. ANTONIADI: *B.S.A.F.*, Vol. 11, p. 487 (1897).

11. J. I. PLUMMER: *Month. Not. R.A.S.*, Vol. 36, p. 351 (1876).

12. W. H. STEAVENSON: *J.B.A.A.*, Vol. 66, p. 264 (1956). See also *Sky and Telescope*, Vol. 16, p. 161 (1956).

13. Various authors: *B.S.A.F.*, Vol. 14, p. 236 (1900). See also WEST and BURT, *Marine Observer*, Vol. 15, p. 8 (1938) and *Nature* (1930), p. 141, 168. Also E. M. ANTONIADI, *L'Astronomie*, Vol. 50, p. 182 (1936) and W. GROUBE, *L'Astronomie*, Vol. 48, p. 116 (1934).

14. A. J. L. MURRAY, *New Scientist*, Vol. 4, p. 133 (5 June 1958) and also personal correspondence with Moore.

15. For a list of old determinations of the angular diameter of Venus, see HOUZEAU, *Vade-Mecum de l'Astronomie*, p. 461 (Brussels 1882).

16. T. J. J. SEE, *Astr. Nach.* 3676; see also *J.B.A.A.*, Vol. 11, p. 128 (1900) and *Nature*, Vol. 63, p. 212.

17. A. A. NEFEDJEV: *Bull. Engelhardt obs.*, No. 30, 3 *(1953)*.

18. G. DE VAUCOULEURS: 'Geometric and Photographic Parameters of the Terrestrial Planets', *Icarus*, Vol. 3, No. 3, pp. 187–235 (1964).

19. VIDAL: *Conn. des Temps*, p. 375 (1810). The measurements by Vidal were made in October 1807.

20. TENNANT: *Month. Not. R.A.S.*, Vol. 35, p. 345; see also *The Observatory*, Vol. 7, p. 262. See also V. VENTOSA, *Astr. Nach.* 4633, and *Ciel et Terre*, Vol. 34, p. 113 (1918).

Chapter 3: **The Movements of Venus**

1. Those who are interested in the development of the theory of the movements of Venus may consult references such as: LALANDE, J. J. DE: 'Calcul des inégalités de Vénus par l'attraction de la Terre', *Histoire de l'Académie des Sciences*, p. 309. LAPLACE, P. S. DE: 'Théorie de Vénus' (Paris 1760), *Traité de Mécanique Céleste*, III, liv. IV, ch. 9 (1802). LE VERRIER, U. J. J.: 'Théorie du mouvement de Vénus', *Annales de l'Observatoire de Paris*; *Mémoires*, VI, 1, 169 (1861). LE VERRIER, U. J. J.: *Tables génerales du mouvement de Venus*, Ibid., p. 95. HILL, G.: Tables of Venus, prepared for the use of the American Ephemeris and Nautical Almanac, Washington

(1872). A survey of research into the movements of Venus up to 1811 was given by B. von Lindenau (see F. von Zach: *Monatliche Correspondenz zur Beförderung der Erde-und Himmelskunde*, XXIII, p. 207 (Gotha 1811).

2. Pliny: *Naturalis Historia*, II, 37.

3. Williamson: *Religious and Cosmic Beliefs of Central Polynesia*, I, p. 128.

4. Layard: *Ninevah and its Remains*, 2, p. 356, and *Ninevah and Babylon*, p. 606. See also C. Reinhardt in *Jnl. Roy. Astron. Soc. Canada*, Vol. 23, p. 113.

5. J. Offord: *J.B.A.A.*, Vol. 13, p. 40 (1902) and *Knowledge* (November 1902); see also 'The Deity of the Crescent Venus in Ancient Western Asia', *Journal of the Royal Asiatic Society*, April 1915.

6. W. W. Campbell: *Publ. Astron. Soc. Pacific*, Vol. 28, p. 85.

7. H. McEwen: *J.B.A.A.*, Vol. 6, p. 34 (1895).

8. H. McEwen: *J.B.A.A.*, Vol. 25, p. 33 (1912).

9. R. A. Proctor: *Orbs Around Us*, p. 102 (London 1881).

10. C. Reinhardt: *Jnl. Roy. Astron. Soc. Canada*, Vol. 25, p. 269 (1931).

11. C. Reinhardt: *Jnl. Roy. Astron. Soc. Canada*, Vol. 23, p. 48 (1929).

12. D. Howell: *Jnl. Roy. Astron. Soc. Canada*, Vol. 25, p. 132 (1931).

13. H. W. Cornell: *Jnl. Roy. Astron. Soc. Canada*, Vol. 29, p. 31 (1935).

14. F. W. Wood: *Jnl. Roy. Astron. Soc. Canada*, Vol. 29, p. 119 (1935).

15. M. A. Blagg: *J.B.A.A.*, Vol. 19, p. 218 (1909).

16. T. W. Webb: *Celestial Objects for Common Telescopes*, Vol. 1, p. 63 (1917 edition).

17. W. S. Franks: personal correspondence with Moore, 1932.

18. E. Halley: 'An Account of the Cause of Venus being Seen in Daylight', *Phil. Trans.*, Vol. 4, pp. 300–2 (1721). See also Halley in ibid., p. 466 (1716).

19. I. Velikovsky: *Worlds in Collision* (London and New York 1950).

20. Patrick Moore: *Can You Speak Venusian?*, (Ian Henry Publications, London, and W. W. Norton Inc., New York, 1978). This is a survey of all the scientific eccentrics, including Dr Velikovsky.

21. M. GARDNER: *Fads and Fallacies in the Name of Science*, p. 31 (London and New York 1962).
22. D. GOLDSMITH (editor): *Scientists Confront Velikovsky* (Cornell University Press 1977).

Chapter 4: **Early Telescopic Views of Venus**

1. J. F. W. HERSCHEL: *Outlines of Astronomy*, p. 311–2 (1850).
2. PATRICK MOORE: *Guide to Mars* (London and New York 1979). This gives a summary of early observations of Mars. For a very detailed survey, see C. FLAMMARION: *La Planète Mars*, Vol. I (Paris 1892) and II (Paris 1910), Paris.
3. F. FONTANA: *Novæ coelestium terrestriumque rerum observationes*, p. 92 (Naples 1646).
4. BURATTINI: *Jnl. des Scavans*, p. 287 (1665) and pp. 33, 101 (1666).
5. G. D. CASSINI: *Jnl. des Scavans*, p. 122 (1667). See also *Phil. Trans.* Vol. 2, p. 615, and E. M. ANTONIADI, *L'Astronomie*, Vol. 39, p. 430 (1925).
6. J. J. CASSINI: *Histoire de l'Acadèmie des Sciences, avec les mèmoires de mathématique*, p. 197 (1732); also *Éléments d'Astronomie*, p. 515 (Paris 1740).
7. J. J. CASSINI: *Histoire de l'Académie des Sciences, avec les mèmoires de mathématique*, p. 213 (1732); also *Éléments d'Astronomie*, p. 525 (Paris 1740).
8. F. BIANCHINI: *Hesperi et Phosphori Nova Phænomena* (Rome 1727).
9. M. V. LOMONOSOV: 'The Appearance of Venus on the Sun as observed in the St. Petersburg Academy of Sciences', contained in the collected works of Lomonosov, edited by M. Sukhomlinev (1891–2).
10. O. STRUVE: *Sky and Telescope*, Vol. 13, p. 118 (1954).
11. See, for instance, N. MENSHUTKIN: *Russia's Lomonosov.* (Princeton University Press, 1952) and V. V. SHARONOV, *Lomonosov* (U.S.S.R. Academy of Sciences, Moscow 1960).
12. J. J. DE LALANDE: *Astronomie*, para. 2272.
13. J. SCHRÖTER: *Aphroditographische Fragmente* (Helmstedt 1796).
14. G. DE VAUCOULEURS: *Discovery of the Universe*, p. 125 (London 1957).

180   *References: pages 40 to 42*

15. PATRICK MOORE: 'A Defence of Schröter', *J.B.A.A.*, Vol. 70, p. 363 (1960).

16. J. SCHRÖTER: *Phil. Trans.*, 1792, pp. 309 ff; see also *Aphr. Frag.*, p. 85.

17. GUTHRIE: 'Observation of an Appearance in the Planet Venus', *Month. Not. R.A.S.*, Vol. 14, p. 169 (1842).

18. L. R. GIBBS, 'Annular Phase of Venus', *Science*, Vol. 16, p. 303 (1890).

19. C. S. LYMAN: *American Journal of Science and Arts*, IIIrd series, IX, p. 47 (1875). See also *L'Astronomie* (June 1891) and *J.B.A.A.*, Vol. 1, p. 455 (1891).

20. See, for instance, H. N. RUSSELL and Z. DANIEL, *Nature*, Vol. 76, p. 389, and *Ap. J.*, Vol. 20, p. 69. For a more modern instance see J. C. BARTLETT, *Strolling Astronomer*, Vol. 70 p. 30 (1953).

21. J. SCHRÖTER: *Cythereographische Fragmente* (Erfurt 1792); *Beobachtungen über die sehr beträchlichten Gebirge und Rotation der Venus*, 35 (1792). See also *Aphr. Frag.*, p. 65, and *Phil. Trans.*, p. 117 (1795).

22. J. SCHRÖTER: *Beilage zu der Aphroditographische Fragmente* (Göttingen); also *Monatliche Correspondenz zur Beförderung der Erde- und Himmelskunde* (Von Zach), XXV, p. 366.

23. J. SCHRÖTER: *Phil. Trans.*, p. 336 (1792).

24. DE GOIMPY: *Jnl. des Scavans* (January 1769).

25. MAIRAN: *Mem. Acad.* (1729).

26. J. SCHRÖTER: *Aphr. Frag.* See also H. McEWEN: *J.B.A.A.*, Vol. 48, p. 62 (1937); T. W. WEBB: *Celestial Objects*, Vol. 1, p. 64 (London 1917), and a relevant observation by Schröter in *Phil. Trans.* pp. 133–4 (1795). See also E. M. ANTONIADI: *L'Astronomie*, Vol. 54, p. 260 (1940), who suggests that some such effect is indicated in the *Histoire Celeste* by LE MONNIER (1741).

27. W. BEER and J. H. MÄDLER: *Beiträge zur physichen Kenntnis der himmlischen Körper in Sonnensysteme* (Weimar 1841); Fragments sur les corps célestes du système solaire (Paris 1840). See also MÄDLER: *Populäre Astronomie* (Berlin 1841).

28. F. DI VICO: *Rotazione di Venere sul proprio asse. Memoria intorno a parecchie osservazioni fatte nella specola dell Universitá gregoriana in Collegio Romano*, p. 42 (1839).

29. An excellent account of this is given in Chapter 2 of *The Planets: Some Myths and Realities*, by R. M. BAUM (David & Charles,

Newton Abbot 1973).

30. W. HERSCHEL: *Phil. Trans.*, pp. 201–9 (1793). Reproduced in Herschel's Collected Scientific Papers, pp. 442–51 (published by the Royal Astronomical Society, 1912).

31. J. SCHRÖTER, *Phil. Trans.*, Vol. 85, pp. 120–1 (1795).

32. J. SCHRÖTER, *Aphr. Frag.*, pp. 13–15 (1796).

33. J. SCHRÖTER: *Phil. Trans.*, Vol. 82, pp. 316–37 (1792).

34. W. HERSCHEL: *Phil. Trans.*, Vol. 83, pp. 202–16 (1793).

35. J. SCHRÖTER: *Phil. Trans.*, Vol. 85, pp. 117–56 (1795).

36. J. BREEN, *The Planetary Worlds*, pp. 154–6 (1854).

37. F. PORRO, *J.B.A.A.*, Vol. 3, p. 184 (1893).

38. W. ALEXANDER: *J.B.A.A.*, Vol. 3, p. 233 (1893).

39. W. F. DENNING: *Astronomical Register*, Vol. 11, p. 131 (1874) and *Month. Not. R.A.S.*, Vol. 42, p. 111 (1882).

40. W. F. DENNING: *Telescopic Work for Starlight Evenings*, p. 148 (1891).

41. LOHSE and WIGGLESWORTH, *Month. Not. R.A.S.*, Vol. 47, p. 495 (1886).

42. C. V. ZENGER, *Month. Not. R.A.S.*, Vol. 37, p. 461 (1877).

43. R. LANGDON, *Month. Not. R.A.S.*, Vol. 33, p. 500 (1873).

44. H. PRATT, *Astronomical Register*, Vol. 10, p. 43 (1873).

45. E. L. TROUVELOT, *Bul. Soc. Astr. France*, Vol. 6, p. 81 (1892). See also Trouvelot in ibid. pp. 61–147; *The Observatory*, Vol. 3, p. 416 (1880) and *Comptes Rendus*, Vol. 98, p. 719 (1884).

46. H. McEWEN: *J.B.A.A.*, Vol. 57, p. 143 (1947).

47. F. VON P. GRUITHUISEN. *Nova acta Academiæ naturæ curiosorum: Verhandlungen der Deutschen Akademie der Naturfoscher*, Vol. 10, p. 239 (Bonn 1821).

48. VOGEL and LOHSE: *Bothkamp Beobachtungen*, heft 2, p. 120 (1872–5).

49. G. V. SCHIAPARELLI: *Ciel et Terre*, Vol. 11 (1890–1), a translation of his original paper in *Rendiconti del R. Istituto Lombardo*, 23, series 2 (1890). See also Schiaparelli in *Astr. Nach.*, Vol. 138, p. 249 (1895).

50. B. DE LA GRYE, *Comptes Rendus*, Vol. 98, p. 1406 (1884).

51. P. LOWELL: *The Evolution of Worlds*, p. 77 (1909).

52. L. NIESTEN and E. STUYVAERT, *Observations sur l'Aspect physique de'Vénus de 1881 à 1895* (Brussels 1903).

53. L. BRENNER: *J.B.A.A.*, Vol. 6, p. 45 (1895) and *English Mechanic*, Vol. XLII, pp. 88 and 137; see also *The Observatory*, Vol. 19, p. 161.

54. L. Brenner: *J.B.A.A.*, Vol. 7, p. 281 (1897).
55. G. V. Schiaparelli: *Rendiconti del R. Istituto Lombardo*, Vol. 23; see also *Astr. Nach.* 2944; *Month. Not. R.A.S.*, Vol. 41, p. 246; *Ciel at Terre*, Vol. 11, pp. 49, 125, 183, 214 and 259.
56. A. Dollfus: *L'Astronomie*, Vol. 69, p. 418 (1955).
57. Perrotin: *Nature*, Vol. 43, pp. 22 and 52; see also *Comptes Rendus*, Vol. 111, p. 542, and Vol. 122, p. 395.
58. P. Lowell: *Month. Not. R.A.S.*, Vol. LVII, p. 148 (1897).
59. See, for instance, Lowell's papers in *Astr. Nach.*, No. 3823; *Nature*, Vol. 55, p. 421; Vol. 67, p. 67; Vol. 69, p. 424; and Vol. 82, p. 260; *J.B.A.A.*, Vol. 7, pp. 87, 213 and 315; *Popular Astronomy*, Vol. 4, pp. 281 and 389; Vol. 11, p. 426; and Vol. 12, p. 184; *The Observatory*, Vol. 20, pp. 172 and 208; Vol. 26, p. 398; and Vol. 36, p. 308; *Popular Science*, Vol. LXXXV, p. 521; and various issues of the Lowell Observatory Bulletin. See also Holden in *Publ. Astron. Soc. Pacific*, Vol. 9, p. 92.
60. L. Brenner: *The Observatory* (May 1897) and *J.B.A.A.* Vol. 7, p. 408 (1897).
61. E. M. Antoniadi: *J.B.A.A.*, Vol. 8, p. 45 (1897).
62. A. E. Douglass: 'The Markings on Venus', *Month. Not. R.A.S.*, Vol. 58, pp. 382–5 (1898).
63. R. Barker: *J.B.A.A.*, Vol. 42, p. 216 (1932).
64. R. Barker: *J.B.A.A.*, Vol. 43, p. 159 (1933) and Vol. 44, p. 302 (1934).
65. R. M. Baum: *Urania*, 229 (Barcelona 1952).
66. O. C. Ranck: *Strolling Astronomer*, Vol. 8, p. 66 (1954).
67. J. C. Bartlett: *Strolling Astronomer*, Vol. 9, pp. 2–8, 1955.
68. Patrick Moore: *Strolling Astronomer*, Vol. 9, pp. 50–54 and 112 (1955).
69. R. M. Baum: *Strolling Astronomer*, Vol. 9, p. 82 (1955).
70. J. C. Bartlett: *Strolling Astronomer*, Vol. 10, p. 102 (1956).
71. A. Dollfus: *L'Astronomie*, Vol. 69, p. 413 (1955).
72. K. Brasch, *Strolling Astronomer*, Vol. 15, p. 156 (1961).

Chapter 5: **Venus in the Twentieth Century: Observational Results**

1. H. Spencer Jones: *Life on Other Worlds*, p. 156 (London 1953).
2. W. Villiger: *Neue Annalen*, Vol. 3 (Munich Observatory, 1898).

3. W. W. SPANGENBERG: *Astr. Nach.*, Vol. 281, p. 6 (1952).
4. A. P. LENHAM and J. H. LUDLOW: *J.B.A.A.*, Vol. 64, p. 300 (1954).
5. PATRICK MOORE: *Sky and Telescope*, Vol. 17, p. 179 (1958).
6. PATRICK MOORE and P. J. CATTERMOLE: *J.B.A.A.*, Vol. 70, p. 130 (1960).
7. E. E. BARNARD: *Astrophysical Journal*, p. 299 (1897).
8. E. M. ANTONIADI: *L'Astronomie*, p. 210 (1928).
9. A. P. LENHAM: *J.B.A.A.*, Vol. 68, p. 98 (1958).
10. W. H. STEAVENSON: *J.B.A.A.*, Vol. 34, p. 126 (1924).
11. W. H. PICKERING: *J.B.A.A.*, Vol. 31, p. 218 (1921).
12. H. McEWEN: *J.B.A.A.*, Vol. 36, p. 191 (1925).
13. W. H. STEAVENSON: *J.B.A.A.*, Vol. 36, p. 297 (1925).
14. W. H. STEAVENSON: *J.B.A.A.*, Vol. 36, p. 301 (1925).
15. E. M. ANTONIADI: *J.B.A.A.*, Vol. 39, p. 88 (1929).
16. J. CAMUS: *L'Astronomie*, Vol. 46, p. 145, 1932. See also *J.B.A.A.*, Vol. 42, p. 311 (1932).
17. L. ANDRENKO: *Urania*, 110 (Barcelona 25 Nov 1935). See also *L'Astronomie*, Vol. 46, p. 114 (1932) and a note by T. L. MACDONALD in *J.B.A.A.*, Vol. 46, p. 207 (1936).
18. H. McEWEN: *J.B.A.A.*, Vol. 36, p. 190 (1926).
19. A. DOLLFUS: 'Étude visuelle et photographique de l'atmosphère de Vénus', *L'Astronomie*, Vol. 69, p. 415 (1955). Reported in *Sky and Telescope*, Vol. 15, p. 397 (1956).
20. H. C. UREY: 'Les Molécules dans les Astres', communications présentées au septième Colloque International d'Astrophysique tenu à Liège, 12–14 July 1956, p. 160 (Liège 1957).
21. PATRICK MOORE: *The Planet Venus*, p. 50 (London 1961).
22. R. JARRY-DESLOGES: 'Observations des Surfaces Planétaires', *Fascicule* VIII, p. 190 (Années 1921 et 1922); see also McEWEN, *J.B.A.A.*, Vol. 46, p. 145 (1936).
23. H. McEWEN: *J.B.A.A.*, Vol. 46, p. 145 (1936).
24. Changes in the colour of Venus have been reported by N. BARABASHOV, *Publ. Kharkov Obs.*, Vol. 2, 5 (1952). Interesting comments with regard to the colour of Venus have been made by G. RAYMOND in *L'Astronomie*, Vol. 36, p. 135 (1922) and by G. ORIANO in *L'Astronomie*, Vol. 52, p. 400 (1938). F. LINK and L. NEUZIL (*Les Molécules dans les Astres*, pp. 156–9, Liège 1957) stated that Venus is definitely yellower than the Sun. See also D. H. MENZEL and F. L. WHIPPLE, *Publ. Astron. Soc. Pacific*, Vol. 67, p. 161 (1955) and G.

KUIPER, *Atmospheres of the Earth and Planets*, p. 308 (1952).

25. J. A. LEES: *J.B.A.A.*, Vol. 39, p. 342 (1928).

26. 'Report of the Mercury and Venus Section': *J.B.A.A.*, Vol. 67, p. 306 (1957).

27. W. H. STEAVENSON: *J.B.A.A.*, Vol. 36, p. 299 (1926).

28. G. P. KUIPER: *Astrophysical Journal*, November 1954. See also *Sky and Telescope*, Vol. 14, p. 131 (1955).

29. A. DOLLFUS: *L'Astronomie*, Vol. 69, p. 425 (1955).

30. F. BIANCHINI: *Hesperi et Phosphori Nova Phænomena* (Rome 1727).

31. J. SCHRÖTER: *Aph. Frag.*, 1796.

32. W. H. PICKERING: *J.B.A.A.*, Vol. 31, p. 218 (1920).

33. R. JARRY-DESLOGES: 'Observations des Surfaces Planétaires', *Fasc.* (1933).

34. E. M. ANTONIADI: *Astr. Nach.*, 6246 (1934).

35. W. H. HAAS: *Jnl. Roy. Astron. Soc. Canada*, Vol. 37, p. 323 (1942).

36. SCHIRDEWAHN: *Sternwalt*, No. 1 (1950).

37. KUTSCHER: *Bull. Astr. Geod. Soc. U.S.S.R.*, 1954.

38. F. E. ROSS: *Astrophysical Journal*, Vol. 68, pp. 57–92 (July 1928).

39. E. M. ANTONIADI, *J.B.A.A.*, Vol. 44, p. 342 (1934).

40. PATRICK MOORE: *J.B.A.A.*, Vol. 65, p. 235 (1955).

41. W. A. GRANGER: *J.B.A.A.*, Vol. 67, p. 104 (1957).

42. J. HEDLEY ROBINSON: *J.B.A.A.*, Vol. 66, p. 261 (1956).

43. 'Report on the Observation of the Planet Venus': *Memoirs of the B.A.A.*, Vol. 41 (December 1974).

44. PATRICK MOORE: *J.B.A.A.*, Vol. 65, p. 234 (1955).

45. W. H. STEAVENSON: *J.B.A.A.*, Vol. 36, p. 274 (1926).

46. *Bull. Astron. Geod. Soc. U.S.S.R.*, No. 12, 3 (1953).

47. See, for instance, FONTSÉRÉ: *Ciel et Terre*, Vol. 18, p. 354 (1897) and *Astr. Nach.* 3430; E. M. NELSON; *L'Astronomie*, Vol. 33, pp. 328 and 368 (1919); and C. FLAMMARION: *L'Astronomi*, Vol. 33, p. 361 (1919).

48. J.B.A.A., *37m* 345 (1927).

49. J. BARTLETT, *Strolling Astronomer*, Vol. 7, p. 34 (1953).

50. M. B. B. HEATH: *J.B.A.A.*, Vol. 66, p. 34 (1955).

51. W. SANDNER: *Mitt. für Plan.*, Vol. 10, No. 1 (1957). See also a note by MOORE in *J.B.A.A.*, Vol. 68, p. 39 (1958).

52. V. A. BRONSHTEN: *Planets and their Observation* (Moscow 1957). (In Russian).

53. *J.B.A.A.*, Vol. 67, p. 102 (1956).
54. *J.B.A.A.*, Vol. 69, p. 22 (1957).
55. *J.B.A.A.*, Vol. 71, p. 64 (1959).
56. *J.B.A.A.*, Vol. 72, p. 262 (1961).
57. *J.B.A.A.*, Vol. 73, p. 184 (1962).
58. *J.B.A.A.*, Vol. 77, p. 126 (1964).
59. *J.B.A.A.*, Vol. 77, p. 339 (1965).
60. *J.B.A.A.*, Vol. 78, p. 223 (1967).
61. *J.B.A.A.*, Vol. 80, p. 55 (1969).
62. *J.B.A.A.*, Vol. 81, p. 224 (1970).
63. *J.B.A.A.*, Vol. 83, p. 38 (1972).
64. *J.B.A.A.*, Vol. 69, p. 108 (1958).
65. *J.B.A.A.*, Vol. 71, p. 146 (1959).
66. *J.B.A.A.*, Vol. 73, p. 103 (1961).
67. *J.B.A.A.*, Vol. 77, p. 126 (1963).
68. *J.B.A.A.*, Vol. 77, p. 127 (1965).
69. *J.B.A.A.*, Vol. 77, p. 339 (1966).
70. *J.B.A.A.*, Vol. 79, p. 50 (1967).
71. *J.B.A.A.*, Vol. 80, p. 385 (1969).
72. *J.B.A.A.*, Vol. 82, p. 50 (1971).
73. *J.B.A.A.*, Vol. 83, p. 449 (1972).
74. H. BRINTON and P. MOORE: *J.B.A.A.*, Vol. 73, p. 119 (1963).
75. W. H. WRIGHT: *Publ. Astron. Soc. Pacific*, Vol. 39, p. 220 (1927).
76. F. E. ROSS: *Astrophysical Journal*, Vol. 68, p. 57 (1928). See also *L'Astronomie*, Vol. 43, p. 367 (1929).
77. R. S. RICHARDSON: *Publ. Astron. Soc. Pacific*, Vol. 67, p. 304 (1955).
78. G. P. KUIPER: *Astrophysical Journal*, Vol. 120, p. 603 (1954) and Contributions from the McDonald Observatory of Texas, 246. See also *Sky and Telescope*, Vol. 14, p. 141 (1954).
79. N. A. KOZYREV: *Publ. of the Crimean Astrophysical Observatory*, Vol. 12, p. 177 (1954); also *Sky and Telescope*, Vol. 15, p. 159 (1956).
80. A. DOLLFUS: *L'Astronomie*, Vol. 67, p. 61 (1953) and Vol. 69, p. 413 (1955).
81. R. S. RICHARDSON: *Publ. Astron. Soc. Pacific*, Vol. 67, p. 304 (1955).
82. A. DOLLFUS: *L'Astronomie*, Vol. 69, p. 425 (1955).
83. BOYER and CAMICHEL: *Ann. Astrophys.*, Vol. 24, p. 531 (1961).
84. BOYER and GUÈRIN: *Icarus*, Vol. 11, p. 338 (1966).

Chapter 6: **Venus in the Twentieth Century:**
        Spectroscope and Theory

1. TACCHINI and RICCÒ: *Mem. Sprettr. Italiani* (December 1882).
2. YOUNG: *American Journal of Science, and Arts*, Vol. 35, p. 328.
3. SCHEINER: quoted by RICHARDSON, *Man and the Planets*, p. 150 (London 1954).
4. SLIPHER: *Lowell Observatory Bulletin*, Vol. 3, p. 88 (1921).
5. NICHOLSON and ST. JOHN: *Phys. Rev.*, Vol. 19, p. 444 (1922). See also the address by W. St. John at the Centenary of the Royal Astronomical Society on 30 May 1922, quoted in *J.B.A.A.*, Vol. 32, p. 286 (1922). See also *Astrophysical Journal*, Vol. 56, p. 380 (1922).
6. W. S. ADAMS and T. DUNHAM, *Publ. Astron. Soc. Pacific*, Vol. 44, p. 243 (1932).
7. T. DUNHAM: *Publ. Astron. Soc. Pacific*, Vol. 45, p. 202 (1932).
8. JOHNSTONE STONEY: *Nature*, 30 Dec. 1897, and *Trans. Roy. Dublin Soc.*, November 1897.
9. A. ADEL: *Astrophysical Journal*, Vol. 93, p. 397 (1941).
10. *Summary of the Moore-Ross flight of November 1959*: John Hopkins University, Baltimore (1960).
11. WATSON: *Nature*, Vol. 12, p. 448.
12. H. SPENCER JONES: *Life on Other Worlds*, p. 167 (London 1952).
13. B. WARNER: *Month. Not. R.A.S.*, Vol. 121, p. 279 (1960).
14. V. A. FIRSOFF: *Our Neighbour Worlds*, p. 209 (London 1952).
15. N. BARABASHOV, *Technika Molodezhi*, p. 14 (1960–4).
16. R. WILDT: *Astrophysical Journal*, Vol. 86, p. 321 (1937).
17. R. WILDT: *Astrophysical Journal*, Vol. 92, p. 247 (1940). See also B. M. PEEK, *J.B.A.A.*, Vol. 51, p. 102 (1941).
18. T. DUNHAM: in *Atmospheres of the Earth and Planets* (ed. G. P. Kuiper), p. 296 (1952).
19. Quoted by G. P. KUIPER, *Atmospheres of the Earth and Planets*, p. 371 (1952).
20. A. W. CLAYDEN: *Month. Not. R.A.S.*, Vol. LXIX, p. 95 (1908); see also *J.B.A.A.*, Vol. 19, p. 227 (1909).
21. F. E. ROSS: *Astrophysical Journal*, Vol. 68, p. 57 (1928). See also *L'Astronomie*, Vol. 43, p. 367 (1929) and H. McEWEN in *J.B.A.A.*, Vol. 39, p. 52 (1928).
22. A. DOLLFUS: *L'Astronomie*, Vol. 69, p. 425 (1955).
23. G. P. KUIPER: *Atmospheres of the Earth and Planets*, p. 371 (1952), and *Transactions of the International Astronomical*

Union, p. 251 (1955).

24. P. LOWELL: *The Evolution of Worlds* (New York 1909).
25. B. GERASIMOVIČ: *Pulkovo Bulletin*, No. 127 (1937).
26. C. TOMBAUGH: *Astronomical Journal*, Vol. 55, p. 184 (1950).
27. W. RABE: *Astr. Nach.*, Vol. 276, p. 111 (1948).
28. F. LINK: *Bull. Astr. Inst. Czech.*, Vol. 1, p. 75 (1949).
29. V. V. SHARONOV: *Astr. Circ. U.S.S.R.*, No. 125, pp. 8–9 (1952) and *Astron. Jnl. U.S.S.R.*, Vol. 29, pp. 728–37 (1952). See also V. N. FROLOV: *Bull. Leningrad Obs.*, Vol. 190, p. 62 (1957).
30. A. BELOPOLSKY: *Astr. Nach.*, Vol. 152, p. 263 (1900). See also *Nature*, 14 June 1900; *J.B.A.A.*, Vol. 10, p. 334 (1900); *The Observatory*, Vol. 23, p. 225 (1900) and *Popular Astronomy*, Vol. 8, p. 290 (1900).
31. A. BELOPOLSKY: *Publ. Pulkovo Obs.* (1911); *Comptes Rendus*, Vol. 153, p. 15; see also *J.B.A.A.*, Vol. 22, p. 60 (1911): *The Observatory*, Vol. 34, p. 316; and *Popular Astronomy*, Vol. 19, p. 503.
32. M. W. OVENDEN: *Looking at the Stars*, p. 86 (London 1957).
33. M. W. OVENDEN: Ibid., p. 84.
34. R. S. RICHARDSON: *Publ. Astron. Soc. Pacific*, Vol. 70, pp. 251–60 (1958). See also *Sky and Telescope*, Vol. 17, p. 547 (1958).
35. E. PETTIT and S. B. NICHOLSON: *Publ. Astron. Soc. Pacific*, Vol. 36, p. 227 (1924). See also *L'Astronomie*, Vol. 38, p. 241 (1924) and *Popular Astronomy*, Vol. 32, p. 14 (1924).
36. E. PETTIT and S. B. NICHOLSON: *Publ. Astron. Soc. Pacific*, Vol. 67, p. 293 ff (1955). There is a misprint in this paper (p. 303) where the temperatures of Venus are wrongly given, and a correction is inserted on p. 432 of the same volume. The misprint is repeated by R. S. RICHARDSON in his paper in *Publ. Astron. Soc. Pacific*, Vol. 70, p. 256 (1958).
37. N. A. KOZYREV: *Publ. Crimean Astrophys. Obs.*, Vol. 12, p. 177 (1954). See also *Sky and Telescope*, Vol. 15, p. 159 (1956).
38. W. M. SINTON and J. STRONG: *Science*, Vol. 123, p. 676 (1956).
39. C. E. HOUSDEN: *Is Venus Inhabited?* (Longmans Green, London 1915).
40. *Nature*, Vol. 96, p. 340 (1915).
41. SVANTE ARRHENIUS: *The Destinies of the Stars* (London 1918).
42. F. W. HENKEL: *J.B.A.A.*, Vol. 19, p. 362 (1909).
43. G. ZECH: *J.B.A.A.*, Vol. 32, p. 225 (1922).
44. C. G. ABBOT: *J.B.A.A.*, Vol. 30, p. 198 (1920).

45. F. E. Ross: *Astrophysical Journal*, Vol. 68, p. 57 (1928). See also H. McEwen: *J.B.A.A.*, Vol. 39, p. 52 (1928).
46. A. Adel: *Astrophysical Journal*, Vol. 86, p. 337 (1937).
47. R. Wildt: *Astrophysical Journal*, Vol. 91, p. 226 (1940).
48. G. Herzberg: *Jnl. Roy. Astr. Soc. Canada*, Vol. 45, p. 100 (1951).
49. N. A. Kozyrev: *Publ. Crimean Astrophys. Obs.*, Vol. 12, p. 177 (1954).
50. F. Hoyle: *Frontiers of Astronomy*, pp. 70–1 (London 1955).
51. D. H. Menzel and F. L. Whipple: *Publ. Astron. Soc. Pacific*, Vol. 67, p. 161 ff (1955). See also *Sky and Telescope*, Vol. 14, p. 20 (1954).
52. B. Lyot: *Ann. Obs. Paris (Meudon)*, Vol. 8, p. 766 (1929). See also *L'Astronomie*, Vol. 38, p. 102 (1924) and Vol. 41, p. 275 (1927).
53. W. H. Pickering: *J.B.A.A.*, Vol. 36, p. 303 (1926).
54. H. Jeffreys: *J.B.A.A.*, Vol. 31, p. 307 (1921).
55. H. C. Urey: *The Planets*, pp. 149 and 222 (Oxford 1952).
56. G. A. Tikhov: *J.B.A.A.*, Vol. 65, p. 200 (1955).
57. H. C. Urey: 'Les Molécules dans les Astres', communications présentés au septième Colloque International d'Astrophysique tenu à Liège, 12–14 July 1956, p. 160 (Liège 1957).
58. E. Öpik: *J. Geophys. Res.*, Vol. 66, 2807 (1961).
59. J. D. Kraus: *Nature*, Vol. 178, pp. 33 and 103 (1956).
60. J. D. Kraus: *Nature*, Vol. 178, p. 159 (1956).
61. J. D. Kraus: *Nature*, Vol. 178, p. 687 (1956).
62. C. E. St. John and S. B. Nicholson: 'The physical constituents of the atmosphere of Venus', *Physical Review*, Vol. 19, pp. 444–51 (1922).
63. W. S. Adams and T. Dunham: 'Absorption bands in the infra-red spectrum of Venus', *Publ. Astron. Soc. Pacific*, Vol. 44, pp. 243–9 (1932).
64. E. Pettit and S. B. Nicholson: 'Radiation from the dark hemisphere of Venus', *Publ. Astron. Soc. Pacific*, Vol. 36, p. 227 (1924).
65. E. Pettit and S. B. Nicholson: 'Temperatures on the dark and bright sides of Venus', *Publ. Astron. Soc. Pacific*, Vol. 37, pp. 293–303 (1925).
66. E. M. Goldstein and R. L. Carpenter: 'Rotation of Venus: Period estimated from radar measurements', *Science*, Vol. 139, pp. 910–11 (1963).

67. R. L. CARPENTER: 'A radar determination of the rotation of Venus', *Astron. J.*, Vol. 75, pp. 61–6 (1970).
68. C. BOYER and H. CAMICHEL: 'Observations photographiques de la planète Vénus', *Ann. Astophys.*, Vol. 24, pp. 531–5 (1961).
69. R. F. BEEBE: 'Ultraviolet clouds on Venus; observational bias', *Icarus*, Vol. 17, pp. 602–7 (1972).
70. W. TRAUB and N. P. CARLETON: 'Spectroscopic observations of the Venus clouds', *J. Atmos. Sci.*, Vol. 32, pp. 1045–51 (1975).
71. A. E. E. RODGERS, R. P. INGAILS and L. P. RAINVILLE: 'The topography of a swarth around the equator of the planet Venus made at wavelength dependent on the radar cross section', *Ap. J.*, Vol. 77, 100–8 (1972).
72. D. B. CAMPBELL, R. B. DYCE, R. P. INGALLS, G. H. PETTENGILL and I. I. SHAPIRO: 'Venus: Topography revealed by radar data', *Science*, Vol. 175, pp. 514–15 (1972).
73. J. E. HANSH and A. ARKIG: 'The Clouds of Venus; evidence of their nature', *Science*, Vol. 171, pp. 669–72 (1971).
74. J. E. HANSH and J. W. HOVENIER: 'The interpretation of the polarization of Venus', *J. Atmos. Sci.*, Vol. 31, pp. 1137–60 (1974).
75. G. T. SILL: 'Sulphuric acid in the Venus clouds', *Lunar. Plan. Lab.*, Vol. 171, pp. 191–8 (1972).
76. A. T. YOUNG: 'Are the clouds of Venus sulphuric acid?', *Icarus*, Vol. 18, pp. 564–82 (1973).
77. L. D. G. YOUNG: 'High resolution spectra of Venus—a review', *Icarus*, Vol. 17, pp. 632–58 (1972).
78. C. H. MAYER, T. P. MCCULLOUGH, R. M. SLOANAKER: 'Observations of Venus at 3·15 cm. wavelength', *Ap. J.*, Vol. 127, pp. 1–10 (1958).

Chapter 7: **The Ashen Light**

1. G. RICCIOLI: *Almagestum Novum, Bononiæ* (1651). See also C. V. ZENGER in *Month. Not. Roy. Astron. Soc.*, Vol. 43, p. 331 (1883).
2. R. M. BAUM: *J.B.A.A.*, Vol. 67, p. 242 (1957).
3. *Nature*, Vol. 14, pp. 91 and 131. See also *Astr. Nach.*, No. 1586.
4. A. SAFARIK: *Report of the British Association for the*

Advancement of Science, 1873, pp. 404–8 (Bradford 1873).

5. W. DERHAM: *Astro-theology*, Book V, Chapter 1 (London 1714). Four further editions of the book were issued before 1726.

6. C. KIRCH: *Astr. Nach.*, No. 1586, Vol. LXVII, p. 27 (1726).

7. A. MAYER: *Observationes Veneris Gryphiswaldenses*, p. 19 (1762).

8. F. HAHN: *Berliner astronomisches Jahrbuch 1793*, p. 188 (1793).

9. W. HERSCHEL: 'On the Planet Venus', *Phil. Trans.*, 1793.

10. J. SCHRÖTER: *Berliner astronomiches Jahrbuch fur 1809*, p. 164 (1809) and *Beobachtungen des grossen Cometen von 1807*, Appendix, p. 66.

11. C. HARDING: *Berliner astronomisches Jahrbuch fur 1809*, p. 169.

12. J. W. PASTORFF: *Berliner astronomisches Jahrbuch fur 1825*, p. 235.

13. M. GUTHRIE: *Month. Not. R.A.S.*, Vol. 14, p. 169 (1843).

14. BERRY, *Month. Not. R.A.S.*, Vol. 22, p. 158 (1862).

15. C. L. PRINCE, *Month. Not. R.A.S.*, Vol. 24, p. 25 (1863).

16. W. ENGELMANN: *Astr. Nach.*, No. 1526, Vol. LXIV, p. 223 (1865).

17. T. PETTY, *Astronomical Register*, No. 68, p. 181 (1868).

18. BROWNING: *Astronomical Register*, No. 88, p. 74 and No. 131, p. 281 (1870).

19. A. WINNECKE: *Astr. Nach.* Nos. 1863 and 1866, Vol. 78, pp. 236 and 287 (1871).

20. O. VAN ERTBORN: *Bull. de l'Académie Roy. des Sci. des Lettres des Beaux Arts de Belgique*, 2nd ser., Vol. 43, pp. 20–4 (1877).

21. C. V. ZENGER: *Month. Not. R.A.S.*, Vol. 37, p. 461, supplement (1877).

22. T. W. WEBB: *Astronomical Register*, Vol. 16, p. 76 (1878).

23. C. V. ZENGER: *Month. Not. R.A.S.*, Vol. 43, p. 333 (1883).

24. LOHSE and WIGGLESWORTH: *Month. Not. R.A.S.*, Vol. 47, p. 495 (1887).

25. H. McEWEN: *J.B.A.A.*, Vol. 6, p. 121 (1896).

26. E. E. BARNARD: *Astrophysical Journal, 1897*, p. 299.

27. L. BRENNER: *Astr. Nach.*, No. 3332 (1895).

28. J. C. BARTLETT: *Strolling Astronomer*, Vol. 5, No. 12 (December 1951).

29. C. S. SAXTON: *J.B.A.A.*, Vol. 38, p. 65 (1927).

30. There is a good summary of results up to 1895 by S. M. B. GEMMILL in *J.B.A.A.*, Vol. 5, p. 412 (1895).

31. M. B. B. HEATH: personal correspondence with Moore.
32. Translation by W. LEY, *Rockets, Missiles and Space Travel*, p. 36 (London 1951).
33. J. RHEINAUER: *Die Erleuchtung des Planeten Venus durch die Erde* (Freiburg 1859) and *Grundzuge der Photometrie*, pp. 58–77 (1859).
34. H. KLEIN: *Die Phosphorescenz der Nachtseite der Venus.— Aleitung zur Durchmusterung des Himmels*, p. 101, (Braunschweig 1880).
35. H. VOGEL: *Beobachtungen auf der Sternewarte zu Bothkamp*, Vol. 2, pp. 118–32 (1871).
36. R. BARKER: *J.B.A.A.*, Vol. 64, p. 60 (1954).
37. D. BARBIER: 'La Lumière Cendrée de Vénus', *L'Astronomie*, Vol. 48, pp. 289–96 (1934) and Vol. 50, pp. 27–32 (1936).
38. A DANJON: 'Sur la prétendue Lumière Cendrée de Vénus', *L'Astronomie*, Vol. 48, pp. 370–2 (1934).
39. P. DE HEEN: 'De la lumière secondiare de Vénus', *Bulletin hebdomadaire de l'Association Scientifique de France*, lre sér, Vol. 11, p. 278 (Paris 1872).
40. J. LAMP: *Astr. Nach.*, No. 2818 (1887).
41. J. HOUTGAST: *Nature*, Vol. 175, p. 678 (1955). See also *Sky and Telescope*, Vol. 14, p. 419 (1955).
42. N. A. KOZYREV: *Publ. of the Crimean Astrophysical Observatory*, Vol. 12, p. 169 (in Russian). See also *Sky and Telescope*, Vol. 15, p. 159 (1955).
43. V. FESENKOV and A. I. OPARIN: *The Universe*, p. 217 (Moscow 1954) (in English).
44. B. WARNER: *Month. Not. R.A.S.*, Vol. 121, pp. 279–83 (1960).
45. G. NEWKIRK: 'The Airglow of Venus,' *Planetary and Space Science*, Vol. 1, pp. 32–6 (1959).
46. G. NEWKIRK and J. L. WEINBERG: 'Airglow of Venus: a Re-examination,' *Planetary and Space Science*, Vol. 5, pp. 163–7 (1963).
47. *Memoirs of the British Astronomical Association*, Vol. 41 (Report of the Mercury and Venus Section), p. 32 (1974).

Chapter 8: **The Phantom Satellite**

1. F. FONTANA: *Novæ cœlestium terrestriumque rerum observationes*, tract v (Naples 1646).
2. J. SHORT: *Phil. Trans.*, Vol. 41, p. 646 (1740).
3. MAYER: *Astr. Jahrbuch, 1788*. See also T. W. WEBB in *Nature*, Vol. 14, p. 193.

4. A. SCHEUTEN, *Astr. Jahrbuch, 1778*, p. 186. See also WEBB, op. cit., and W. T. LYNN: *The Observatory*, Vol. 10, p. 73 (1887).
5. J. ASHBROOK, *Sky and Telescope*, Vol. 13, p. 333 (1954). See also WEBB, op. cit., and LYNN, op. cit.
6. BAUDOUIN: *Mémoire sur la découverte du satellite de Vénus, et sur les nouvelles observations qui viennent d'être faites à ce sujet*, (Paris 1761). See also *Dictionnaire de Physique* (Paris 1789).
7. J. LAMBERT: *Mem. Acad. Berlin*, p. 222 (1773).
8. See H. C. SCHJELLERUP, *Copernicus*, Vol. 2, p. 164.
9. M. HELL: *Ephemerides astronomicæ ad meridianum vindobonensem calculis definitæ*, (1766), 37 (Vindobonæ).
10. BOSCOVICH: *Dissertationes quinque ad dipptricam pertinentes* (1767) (Vindobonæ).
11. VON ENDE: *Monatliche Correspondenz zur Beforderung der Erde—und Himmelskunde (Von Zach)*, Vol. 24, p. 494 (1811).
12. BERTRAND: *Jnl. des Savants*, p. 456 (Paris 1875).
13. W. H. SMYTH: *Cycle of Celestial Objects*, Vol. 1, p. 109 (1844).
14. F. SCHORR: *Der Venusmond* (Braunschweig 1875). See also WEBB (op. cit.) and ASHBROOK (op. cit.)
15. J. HOUZEAU: *Ciel et Terre* (15 May 1884) and *The Observatory*, Vol. 7, p. 222 (1884).
16. P. STROOBANT: *Mem. de l'Acad. de Bruxelles*, Vol. XLIX, No. 5 (1887). See also *Astr. Nach.*, No. 2809 (1887); *Nature*, Vol. 35, pp. 503 and 543; and *Ciel et Terre*, Vol. 8, p. 332 (1887).
17. E. E. BARNARD: *Astr. Nach.* Vol. 172, pp. 25 and 207 (1906); and ibid. Vol. 173, p. 315 (1907).
18. J. ASHBROOK: *Sky and Telescope*, Vol. 15, p. 356 (1956).

Chapter 9: **Transits and Occultations**

1. J. MEEUS: *J.B.A.A.*, Vol. 68, p. 98 (1958).
2. C. TOMBAUGH: *Astronomical Journal*, Vol. 55, p. 184 (1950).
3. J. KEPLER: *Admonitio ad Astronomos rerumque Coelestium Studiosos de Miris Rarisque anni 1631 Phaenomenis, Veneris puta at Mercurii in Solem incursu*, (Lipsiæ 1629).
4. GASSENDI: *Mercurius in Sole visus* (Paris 1632), reproduced in Gassendi's *Opera omnia*, IV (Lugduni 1658).
5. WHATTON: *Memoir of Horrocks*, p. 109.
6. L. A. SEDILLOT: *Prolégomènes des tables d'Ouloug Beg*, I, p. xviii (Paris 1849).

7. See HORROCKS: *Venus in Sole visa*, contained in HEVELUS, *Mercurius in Sole visus* (Gedani 1662). A very good account of Horrocks' transit study has been given by S. B. GAYTHORPE: *J.B.A.A.*, Vol. 47, p. 60 (1936).

8. HALLEY: see *Phil. Trans.* p. 511 (1691); also W. T. LYNN, *The Observatory*, Vol. 5, p. 175 (1882).

9. J. ENCKE. *Der Venusdurchgang von 1769* (Gotha 1824). Various other values were derived from these two transits; see, for instance, HORNSBY (*Phil. Trans.* 1763, p. 494); SHORT (*Phil. Trans.* 1763, p. 340); PINGRÉ (*Hist. de l'Acad. des Sciences, avec les mém. de math. et de physique*, 1765, p. 32, and 1772, p. 419); AUDIFFREDI (*Investigatio parallaxis Solis*, Rome 1765); PLANMAN (*Phil. Trans.* 1768, p. 127); EULER (*Novi commentarii Academiæ scientarum Petropolitanæ*, Vol. 14, II, 1770, p. 518, Petropolis); HORNSBY (*Phil. Trans.* 1771, p. 579); LALANDE (*Hist. de l'Académie des Sciences, avec les mém. de math. and de physique*, 1771, p. 798); WILLIAMSON (*Trans. of the American Phil. Soc.*, Philadelphia, Vol. 1, 1789, p. 71); PLANMAN (*Abhandlungen der Swedischen Akademie der Wissenschaften*, Leipzig 1772, pp. 183, 358); LEXELL (*Novi commentarii Academiæ scientarum Petropolitanæ*, Vol. 17, p. 609, 1773); HELL (*De parallaxi Solis ex observationibus transitus Veneris anno 1769*, Vienna 1773); DUSEJOUR (*Hist. de l'Academie des Sciences avec les měm. de math. et de physique*, 1781, p. 330, and 1783, p. 289); WALLOT (*Phil. Trans.* 1784, p. 328); LAPLACE (*Conn. des Temps*, Vol. 12, p. 496, 1804); DELAMBRE (*Astronomie*, III, 506, and I, xliv); an earlier estimate by ENCKE (*Die Entfernung der Sonne von der Erde*, Gotha 1822) and POWALKY (*Neue Untersuchungen des Venusdurchgang von 1769*, Kiel 1864) also *Conn. des Temps*, 1867, p. 22).

10. For a really detailed account of the 1761 and 1769 transits, see H. WOOLF: *The Transits of Venus* (Princeton University Press 1959).

11. In the library of King's College, London, there are some manuscript notebooks relating to Kew Observatory, and one of these contains a paper on the subject, 'Observations of the Transit of Venus, 1769, Richmond Observatory'. See also R. S. WHIPPLE, *Proceedings of the Optical Convention of 1926*, part 2, pp. 511–4.

12. See *The Journal of H.M.S. Endeavour*, Genesis Publications

(Guildford 1977). Cook's log is reproduced in facsimile.
13. See H. C. SORBY, *Nature*, Vol. 10, p. 148.
14. References to the 1874 and 1882 transits are so numerous that to list them all would be difficult. Two books by R. A. PROCTOR, *Transits of Venus* and *Studies of Venus Transits in 1874 and 1882* are valuable sources. There are notes in the *Month. Not. R.A.S.*: Vol. 24, p. 173 (Airy); Vol. 28, p. 255 (Stone); Vol. 29, pp. 33, 43, 45–8, 210–11, 249–50, 305–6, 332 and elsewhere; and Vol. 35, p. 345 (Tennant). There are numerous references in *Nature*, as follows:

Vol	Page(s)
5	177, 370
6	69, 81, 110, 228, 423, 460, 494
7	109, 129, 169, 271, 371, 431, 451
9	117, 183, 230, 350, 389, 403, 447, 452, 487
10	11, 27, 33, 49, 66, 73, 86, 114, 151, 158, 172, 190, 426, 449
12	256
15	48
16	144
17	1, 69, 392, 507
18	221 (photographs)
23	231, 388
24	41
25	137, 242, 493, 505
26	102, 185, 223, 269, 329, 352, 446, 584, 636
27	112–4, 132, 154–9, 177, 179, 180, 197, 208, 246, 253, 266, 284, 483, 539, 541
28	90, 377
31	254
37	253
38	600
39	87
48	447.

In the periodical *The Observatory*:

1	148, 150, 166, 290
5	59, 171, 237, 300, 313, 315, 378
6	16, 18, 24, 57, 59, 93, 188, 194, 305
7	81, 212
8	134
10	301, 366
11	132

12   123

16   367

See also *Die Venus-Durchgänge, 1874, 1882*, 5 vols (Berlin 1887–98); F SCHORR: *Der Vorübergang der Venus von der Sonnenschribe* (Braunschweig 1873); A. AUWERS: *Bericht über die Beobachtung des Venus-Durchgangs* (Berlin 1878); G. AIRY (Editor): *Account of the Observations of the Transit of Venus, 8 December 1874* (London 1881); S. NEWCOMB, *Observations of the Transit of Venus, 8–9 December 1874* (Washington 1880).

15. C. WHITMELL: *J.B.A.A.*, Vol. 24, p. 170 (1913).
16. *J.B.A.A.*, Vol. 44, p. 253 (1934).
17. See J. ASHBROOK: *Sky and Telescope*, Vol. 16, p. 68 (1956).
18. J. BEVIS: *Phil. Trans.*, XLI, p. 630, in Latin. See also A. C. D. CROMMELIN, *J.B.A.A.*, Vol. 7, p. 66 (1896) and ASHBROOK (op. cit.).
19. SIMONELLI: *Scientiæ Eclipsium*, IV, 147.
20. J. J. CASSINI: *Hist. de l'Acad. Roy.*, p. 383 (Paris 1737).
21. See also S. J. JOHNSON, *J.B.A.A.*, Vol. 7, p. 142 (1896).
22. I. YAMAMOTO: *J.B.A.A.*, Vol. 38, p. 121 (1928).
23. *Bul. Astr. Soc. France*, September 1910; *Astr. Nach.*, Vol. 185, p. 303 (1910); *J.B.A.A.*, Vol. 21, p. 64 (1910).
24. THIELE: *Publ. Astron. Soc. Pacific*, Vol. 30, p. 166 (1918).
25. W. H. HAAS: *Strolling Astronomer*, Vol. 2, No. 5 (1948). See also J. G. PORTER, *J.B.A.A.* Vol. 58, p. 226 (1948).
26. D. H. MENZEL and G. DE VAUCOULEURS: *Astronomical Journal*, Vol. 65, p. 351 (1960).
27. Interesting notes on the AD 885 conjunction of Venus and Regulus are given by DELAMBRE, *Astronomie due Moyen Âge*, p. 76. See also J. R. HIND in *Month. Not. R.A.S.*, Vol. 34, p. 105 (1874).

Chapter 10: **Rockets to Venus**

1. See C. A. CROSS and P. MOORE: *Atlas of Mercury* (Mitchell Beazley, London 1977).
2. K. YA. KONDRATYEV and G. E. HUNT: *Weather and Climate of the Planets* (Pergamon Press, 1981).
3. A. P. VINOGRADOV, YU. S. SURKHOV and C. P. FLORENSKY: 'The chemical composition of the Venus atmosphere based on the date of the interplanetary station Venera 4', *J. Atmos. Sci.*, Vol. 25, pp. 535–6 (1968).

4. V. S. AVDNEVSKY, M. YA. MAROV and M. K. ROZHDEST-VENSKY: 'A tentative model of the Venus atmosphere based on the measurements of Veneras 5 and 6', *J. Atmos. Sci.*, Vol. 27, pp. 561–5.

5. A. P. VINOGRADOV, YU. A. SURKHOV and B. M. ANDREICHIKOV: 'Investigation of the composition of Venus' atmosphere on automatic space-probes Venera 5 and 6', *Soviet Physics 'Doklady'*, Vol. 15, pp. 4–6 (1970).

6. V. S. AVDNEVSKY, M. YA. MAROV, M. K. ROZHDESTVEN-SKY, N. F. BORODIN and V. V. KERZHANVOICH: 'The soft landing of Venera 7 on the Venus surface and preliminary results of the investigation of the Venus atmosphere', *J. Atmos. Sci.*, Vol. 28, pp. 263–9 (1971).

7. M. YA. MAROV, V. S. AVDNEVSKY, N. F. BORODIN, A. P. EKONOMOV, V. V. KERZHANOVICH, V. P. LYSOV, B. YA. MOSHKIN, M. K. ROZHDESTVENSKY and O. L. RYABOV: 'Preliminary results on the Venus atmosphere from the Venera 8 descent module', *Icarus*, Vol. 20, pp. 417–21 (1973).

8. M. V. KELDYSH: 'Venus exploration with the Venera 9 and 10 spacecraft', *Icarus*, Vol. 30, pp. 605–25 (1977).

9. B. C. MURRAY, M. J. S. BELTON, G. E. DANIELSON, M. E. DAVIES, D. GAULT, B. HAPKE, B. O'LEARY, R. STROM, V. SUOMI and N. TRASK: 'Venus' atmospheric motion and structure from Mariner 10 pictures', *Science*, Vol. 183, pp. 1307–15 (1974).

10. L. COLIN: 'The Pioneer Venus Programme', *J. Geophys. Res.*, Vol. 85, pp. 7578–98 (1980).

Chapter 11: **The Atmosphere of Venus**

1. J. H. HOFFMAN, R. R. HODGES, T. M. DONAHUE and M. B. McELROY: *J. Geophys. Res.*, Vol. 85, pp. 7882–90 (1980).

2. J. B. POLLACK and D. C. BLACK: *Science*, Vol. 205, pp. 56–9 (1979).

3. G. SCHUBERT *et al*: *J. Geophys. Res.*, Vol. 85, pp. 7963–8006 (1980).

4. F. W. TAYLOR *et al*: *J. Geophys. Res.*, Vol. 85, pp. 8007–25 (1980).

5. K. YA. KONDRATYEV and G. E. HUNT: *Weather and Climate of the Planets* (Pergamon Press, 1981).

6. J. B. POLLACK, O. B. TOON and R. BOESE: *J. Geophys. Res.*, Vol. 85, pp. 8223–31 (1980).
7. R. KRALLENBERG et al.: *J. Geophys. Res.*, Vol. 85, pp. 8059–81 (1980).
8. R. KRALLENBERG and D. M. HUNTEN: *J. Geophys. Res.*, Vol. 85, pp. 8039–58 (1980).
9. L. V. KSANFOMALITIC et al.: *Acad. Sci. USSR Space Res. Inst.*, Vol. 5 (1979).
10. W. W. TAYLOR et al.: *Science*, Vol. 205, pp. 112–14 (1979).
11. L. W. ESPOSITO: *J. Geophys. Res.*, Vol. 85, pp. 8151–7 (1980).
12. J. B. POLLACK et al.: *J. Geophys. Res.*, Vol. 85, pp. 8141–50 (1980).
13. W. ROSSOW et al.: *J. Geophys. Res.*, Vol. 85, pp. 8107–28 (1980).
14. B. C. MURRAY et al.: *Science*, Vol. 183, pp. 1307–15 (1974).
15. M. TOMASKO et al.: *J. Geophys. Res.*, Vol. 85, 8167–86 (1980).
16. W. ROSSOW and G. WILLIAMS: *J. Atmos. Sci.*, Vol. 36, pp. 377–89 (1974).
17. M. J. S. BELTON et al.: *J. Atmos. Sci.*, Vol. 33, pp. 1383–93 (1976).
18. M. J. S. BELTON et al.: *J. Atmos. Sci.*, Vol. 33, pp. 1394–1417 (1976).

Chapter 12: **The Surface of Venus**

1. E. M. ANTONIADI: *La Planète Mercure* (Paris 1934); English translation as *The Planet Mercury* (David & Charles, Newton Abbot 1974). See also C. A. CROSS and P. MOORE, *Atlas of Mercury* (Mitchell Beazley, London 1979).
2. G. H. PETTENGILL, D. B. CAMPBELL and H. MASURSKY: *Sci. Amer.* Vol. 243, pp. 54–65 (1980).
3. H. MASURSKY et al.: *J. Geophys. Res.*, Vol. 85, pp. 8232–60 (1980).
4. G. H. PETTENGILL et al.: *J. Geophys. Res.*, Vol. 85, pp. 8261–70 (1980).
5. D. B. CAMPBELL and B. A. BURNS: *J. Geophys. Res.*, Vol. 85, pp. 8271–81 (1980).
6. H. C. RUMSEY, G. A. MORRIS, R. R. GREEN and R. M. GOLDSTEIN: *Icarus*, Vol. 23, pp. 1–14 (1974).
7. C. P. FLORENSKY et al.: *Geol. Soc. Amer. Bull.*, Vol. 88, pp. 1537–45 (1977).

8. G. SCHUBERT *et al.*: *J. Geophys. Res.*, 8007–25 (1980).
9. A. P. VINOGRADOV, A. YU. SURKOV and F. F. KIRNOZOR: *Icarus*, Vol. 20, pp. 253–9 (1973).
10. W. L. SJÖGREN, R. J. PHILIPS, P. W. BIRKELAND and R. N. WIMBERLEY: *J. Geophys. Res.*, Vol. 85, pp. 8295–8302 (1980).

Chapter 13: **The Interior of Venus**

1. H. MASURSKY *et al.*: *J. Geophys. Res.*, Vol 85, pp. 8232–60 (1980).
2. W. L. SJÖGREN *et al.*: *J. Geophys. Res.*, Vol. 85, pp. 8295–303 (1980).
3. C. P. FLORENSKY *et al.*: *Geol. Soc. Amer. Bull.*, Vol. 88, pp. 1537–45 (1977).
4. C. T. RUSSELL, R. C. ELPHIC and J. A. SLAVIN: *J. Geophys. Res.*, Vol. 85, 8319–32 (1980).

Appendix 2: **Estimated Rotation Periods**

1. G. D. CASSINI: *Jnl. des Scavans*, December 1667, p. 122 (p. 182 in the original printing). The article is reproduced in *Histoire de l'Académie des Sciences*, p. 467 (Paris 1731). See also *Nature*, Vol. 13, p. 512.
2. F. BIANCHINI: *Hesperi et Phosphori Nova Phænomena* (Rome 1727). See also *Astr. Nach.*, No. 278, and *Nature*, Vol. 13, p. 512.
3. J. J. CASSINI, *Histoire de l'Académie des Sciences, 1732*, p. 197; also *Élements d'Astronomie*, p. 515 (Paris 1740).
4. J. J. CASSINI: *Histoire de l'Académie des Sciences, 1732*, p. 213; also *Élements d'Astronomie*, p. 525 (Paris 1740).
5. J. SCHRÖTER: *Cythereographiscie Fragmente* (Erfurt 1792); *Beobachtungen über die sehr beträchlichten Gebirge und Rotation der Venus*, p. 35 (1792). See also *Aphr. Frag.*, p. 65, and *Phil. Trans.* 1795, p. 117.
6. FRITSCH: *Astr. Jahrbuch 1804*, p. 214.
7. J. SCHRÖTER: *Beitrage zu der Aphroditographische Fragmente (Göttingen)*; also *Monatliche Correspondenz zur Beförderung der Erde-und Himmelskunde (Von Zach)*, Vol. 25, p. 366.
8. T. HUSSEY: *Month. Not. R.A.S.*, Vol. 2, p. 78 (1832).
9. DI VICO: *Memoriæ intorno a parecchie osservazioni fatte nella specila dell' Università gregoriana in Collegio Romano*, 1840–1, p. 32; 1843, p. 31; 1849, p. 29; 1850, p. 140. See also *Astr.*

Nach., No. 404. For comments on this and other early estimates see also *Nature*, Vol. 13, p. 512.

10. W. F. DENNING: *Month. Not. R.A.S.*, Vol. XLII, p. 109. See also *Jnl. Roy. Astr. Soc. Canada*, Vol. 9, p. 285.
11. G. V. SCHIAPARELLI: *Rendiconti del R. Istituto Lombardo*, Vol. 23; see also *Astr. Nach.*, No. 2944; *Month. Not. R.A.S.*, Vol. LI, p. 246; *Ciel et Terre*, Vol. 11, pp. 49, 125, 183, 214 and 259.
12. PERROTIN: *Comptes Rendus*, Vol. 111, p. 542 and Vol. 122, p. 395. See also *Bull. de la Soc. Belg. d'Ast.*, Vol. 1, p. 109 (1896).
13. TERBY: *Bull. Belg. Acad. Sci.*, 1891, p. 12. See also *J.B.A.A.*, Vol. 1, p. 283.
14. NIESTEN & STUYVAERT: *Bull. Belg. Acad. Sci.*, Vol. 21, p. 452 (1891). See also *J.B.A.A.*, Vol. 2, p. 16; *The Observatory*, Vol. 14, p. 290; *Ciel et Terre*, Vol. 12, p. 217 (1891).
15. LÖSCHART: *Nature*, Vol. 45, p. 210. See also *J.B.A.A.*, Vol. 2, p. 202.
16. E. TROUVELOT: 'Obs. sur les Planètes Vénus et Mercure', *Bull. Soc. Astr. France*, Vol. 6, p. 61 (1892). See also *J.B.A.A.*, Vol. 2, p. 402.
17. C. FLAMMARION: *Comptes Rendus*, Vol. 17, p. 354 (1894). See also *J.B.A.A.*, Vol. 5, p. 133, and *The Observatory*, Vol. 17, p. 354.
18. CERULLI: *Astr. Nach.*, No. 3329.
19. MASCARI: *Astr. Nach.*, No. 3329; *Astrophysical Jnl.*, Vol. 3, p. 226. See also *J.B.A.A.*, Vol. 6, p. 229.
20. G. V. SCHIAPARELLI: *Astr. Nach.*, No. 3304.
21. TACCHINI: *Mem. Sprettr. Italiani*, Vol. 25, p. 93; *Nature*, Vol. LIII, p. 306.
22. A. STANLEY WILLIAMS: *Astr. Nach.*, No. 3300.
23. W. VILLIGER: *Astr. Nach.*, No. 3332.
24. L. BRENNER: *J.B.A.A.*, Vol. 6, p. 45 (1895) and *English Mechanic*, Vol. XLII, pp. 88 and 137; see also *The Observatory*, Vol. 19, p. 161.
25. L. BRENNER: *J.B.A.A.*, Vol. 7, p. 281 (1897).
26. H. MCEWEN: *J.B.A.A.*, Vol. 22, p. 147.
27. FONTSÉRÉ: *Astr. Nach.*, No. 143, p. 357.
28. L. RUDAUX and G. FOURNIER: *Bull. Soc. Astr. de France*, Vol. 13.
29. MÜLLER: see H. Klein, *Jahrbuch der Astronomie* for 1900.
30. A. BELOPOLSKY: *Astr. Nach.*, No. 152, p. 263 (No. 3641); see

also *Nature*, 14 June 1900; *J.B.A.A.*, Vol. 10, p. 334 (1900); *The Observatory*, Vol. 23, p. 225; *Popular Astronomy*, Vol. 8, p. 290.

31. VASSILLIEFF: *Mem. Imp. Acad. Sci. St. Petersburg*, Vol. 11, p.2.
32. ARENDT: *Astr. Nach.*, No. 3803, p. 175. See also *J.B.A.A.*, Vol. 13, p. 41.
33. E. C. SLIPHER: *Astr. Nach.*, Nos. 3891–2 (1903); *Lowell Observatory Bulletin*, Vol. 1, p. 9 (1903) (No. 3). See also *J.B.A.A.*, Vol. 14, p. 41 (1903).
34. STEFANIK: *Jnl. Roy. Astr. Soc. Canada*, Vol. 11, p. 7.
35. HARG: *Gaz. Astr.*, No. 3. See also *J.B.A.A.*, Vol. 18, p. 261.
36. P. LOWELL. See, for instance, Lowell's papers in *Astr. Nach.*, No. 3823; *Nature*, Vol. 55, p. 421; Vol. 67, p. 67; Vol. 69, p. 424 and Vol. 82, p. 260; *J.B.A.A.*, Vol. 7, pp. 87, 213 and 315; *Popular Astronomy*, Vol. 14, pp. 281 and 389; Vol. 11, p. 426 and Vol. 12, p. 184; *The Observatory*, Vol. 20, pp. 172 and 208; Vol. 26, pp. 398 and Vol. 36, p. 308; *Popular Science*, Vol. LXXXV, p. 521; and various papers in the *Lowell Observatory Bulletin*. See also HOLDEN in *Publ. Astron. Soc. Pacific*, Vol. 9, p. 92.
37. SCHOY: *Gaea*, No. 45.
38. PERQUERIAUX: *Bull. Soc. Astr. de France*, Vol. 23.
39. A. BELOPOLSKY: *Publ. Pulkova Obs. 1911*; *Comptes Rendus*, Vol. 153, p. 15; see also *J.B.A.A.*, Vol. 22, p. 60 (1911); *The Observatory*, Vol. 34, p. 316; *Popular Astronomy*, Vol. 19, p. 503.
40. BELLOT: *Comptes Rendus*, Vol. 153.
41. S. BOLTON: *J.B.A.A.*, Vol. 22, p. 145.
42. MAJERT: *Mitt. der V.A.P.*, Vol. 23.
43. W. RABE: *Astr. Nach.*, Bd. 200.
44. S. MAXWELL: *J.B.A.A.*, Vol. 26, p. 158.
45. W. F. A. ELLISON: *J.B.A.A.*, Vol. 26, p. 209.
46. C. E. HOUSDEN: *J.B.A.A.*, Vol. 26, p. 270.
47. D. H. WILSON: *Popular Astronomy*, Vol. 24.
48. H. E. LAU: *Astr. Nach.*, 205.
49. EVERSHED: *Kodaikanal Report*, 1919. See also *J.B.A.A.*, Vol. 30, p. 229.
50. R. JARRY-DESLOGES: *Obs. des Surfaces Planétaires*, 1933, Fasc. 8.
51. W. H. PICKERING: *J.B.A.A.*, Vol. 31, p. 218 (1921).
52. H. KAUL: *Physikalische Zeitschrift*, 15 April 1922. See also

J.B.A.A., Vol. 32, p. 323.

53. See W. SANDNER in *Vega*, Vol. 1, p. 37.

54. J. G. YANES: *Ley acera des rotaciones planetarias*, 1922.

55. A. RORDAME: *J.B.A.A.*, Vol. 33, p. 59; *Popular Astronomy*, Vol. 30, p. 137; *Nature*, Vol. 109, p. 592 (1922).

56. H. McEWEN: *J.B.A.A.*, Vol. 36, p. 191 (1925).

57. W. H. STEAVENSON: *J.B.A.A.*, Vol. 36, p. 297 (1925).

58. A. FOCK: *Astr. Nach.*, No. 221, p. 95.

59. NICHOLSON and ST. JOHN: *J.B.A.A.*, Vol. 35, p. 263. See also W. S. ADAMS, *Carnegie Inst. (Washington) Yearbook*, Vol. 22, p. 192 (1924); also *Astrophysical Jnl.*, Vol. 56, p. 380.

60. F. E. ROSS: *Astrophysical Jnl.*, Vol. 68, p. 57 (1927); see also *J.B.A.A.*, Vol. 38, p. 103 (1927) and *Popular Astronomy*, November 1927, p. 492.

61. N. P. SCHANIN: *Mirovdenie*, 160: *Popular Astronomy*, Vol. 36, p. 49.

62. N. P. SCHANIN: *Mirovdenie*, 170: *Popular Astronomy*, Vol. 36, p. 565.

63. R. JARRY-DESLOGES: *Comptes Rendus*, CLXXXVII.

64. N. P. SANJUTIN: *L'Astronomie*, Vol. XLVI, p. 145 (1929).

65. J. CAMUS: 'Observations of Venus made in 1932', reported in *L'Astronomie*, Vol. 46, p. 145. See also *J.B.A.A.*, Vol. 42, p. 311 (1932).

66. R. BARKER: *J.B.A.A.*, Vol. 42, p. 216 (1932).

67. R. BARKER: *J.B.A.A.*, Vol. 43, p. 159 (1933) and Vol. 44, p. 302 (1934).

68. E. P. MARTZ: *Himmelswelt*, Vol. 44, p. 62; *Popular Astronomy*, Vol. 41, p. 528.

69. E. M. ANTONIADI: *J.B.A.A.*, Vol. 44, p. 341 (1934).

70. L. ANDRENKO: *Urania*, 110 (Barcelona 25 Nov. 1935). See also a note by T. L. MACDONALD, *J.B.A.A.*, Vol. 46, p. 207 (1936). See also *L'Astronomie*, Vol. 46, p. 114 (1932).

71. W. W. SPANGENBERG: *Astr. Nach.*, Bd. 252 and 261.

72. W. H. HAAS: *Popular Astronomy*, Vol. XLVII, p. 574.

73. L. ROMANI: *L'Astronomie*, Vol. LIV.

74. W. H. HAAS: *Jnl. Roy. Astron. Soc. Canada*, Vol. 37, p. 323 (1942).

75. PHOCAS: *Trans. International Astronomical Union*, Vol. 7, p. 161 (1950).

76. SCHIRDEWAHN & SCHWARTZ: *Die Sternwelt*, No. 1 (1950).

77. V. V. VOLKOV: *Bull. Astron. Geod. Soc. U.S.S.R.*, 1949.

78. R. M. Baum: *Urania*, 229 (Barcelona 1952).
79. H. Le Vaux: *Strolling Astronomer*, Vol. 5, No. 6 (1951).
80. J. C. Bartlett: *Strolling Astronomer*, Vol. 6, No. 2 (1952).
81. G. D. Roth: *Die Sterne*, 1953, p. 163.
82. M. Kutscher: *Mitteilung der Planetenbeobachter*, 1954.
83. G. P. Kuiper: *Astrophysical Jnl.* (November 1954). See also *Sky and Telescope*, Vol. 14, p. 131.
84. A. Dollfus: *L'Astronomie*, Vol. 69, p. 418 (1955).
85. J. D. Kraus: *Nature*, Vol. 178, p. 687 (1956).
86. R. S. Richardson: *Publ. Astron. Soc. Pacific*, Vol. 70, pp. 251–60 (1958). See also *Sky and Telescope*, Vol. 17, p. 547 (1958).
87. I. I. Gusev: *Bull. Astr. Geodetic Soc. U.S.S.R.*, No. 23 (30) (1958). This is in Russian. An English translation appeared in *Stardust*, Vol. 3, p. 47 (1960), the bulletin of the Belfast Centre of the Irish Astronomical Society.
88. R. L. Carpenter: *Astron. Jnl.*, Vol. 69, No. 1, pp. 2–11 (1964).
89. R. M. Goldstein: *Astron. Jnl.*, Vol. 69, No. 1, pp. 12–18 (1964).
90. W. K. Klemperer, G. R. Ochs and K. L. Bowles: *Astron. Jnl.* Vol. 69, No. 1, pp. 22–8 (1964).
91. D. O. Muhleman: *Astron. Jnl.* Vol. 69, No. 1, pp. 33–41 (1964).
92. F. D. Drake: *Astron. Jnl.*, Vol. 69, No. 1, pp. 62–4 (1964).
93. O. N. Rzhiga: 'Radar Observations of Venus in the Soviet Union in 1962 (Pt. III, 8)' *Life Sciences and Space Research II*, ed. M. Florkin and A. Dollfus (Amsterdam, 1964).
94. J. E. Ponsonby, J. H. Thomson and K. Imrie: *Nature*, Vol. 204, pp. 63–4 (1964).
95. I. I. Shapiro: *Sky and Telescope*, Vol. 28, p. 341 (Dec. 1964).
96. R. L. Carpenter: *Handbook of the Planet Venus* (L. Koenig, P. W. Murray, C. M. Michaux and H. A. Hyatt, NASA 1967, p. 20.
97. I. I. Shapiro: *Science*, Vol. 157, p. 423 (1967).
98. I. I. Shapiro: *Astrophysical Letters* (1 June 1979).

Subject Index

Name Index

206 *Name Index*